DESIGN FOR MANUFACTURE
Strategies, Principles and Techniques

Addison-Wesley Series in Manufacturing Systems

Consulting editor
Professor J. Browne, University College Galway

DESIGN FOR MANUFACTURE
Strategies, Principles and Techniques

John Corbett
University of Wisconsin

Mike Dooner
University of Hull

John Meleka
The Open University

Christopher Pym
The Open University

ADDISON-WESLEY
PUBLISHING
COMPANY

Wokingham, England · Reading, Massachusetts · Menlo Park, California
New York · Don Mills, Ontario · Amsterdam · Bonn
Sydney · Singapore · Tokyo · Madrid · San Juan · Milan · Paris

Cover designed by Crayon Design of Henley-on-Thames and printed by The Riverside Printing Co. (Reading) Ltd.
Typeset by Keyset Composition, Colchester.
Printed in Great Britain by The Bath Press, Avon.

First printed 1991

British Library Cataloguing in Publication Data
Design for manufacture.
  1. Manufactured goods. Design
  I. Corbett, John
  670

  ISBN 0-201-41694-8

Library of Congress Cataloging in Publication Data
Design for manufacture : strategies, principles, and techniques /
  edited by John Corbett . . . [et al.].
    p. cm.
  Includes bibliographical references and index.
  ISBN 0-201-41694-8
  1. Engineering design.   2. Production engineering.   I. Corbett,
John, 1940–
TA174.D467   1991
670—dc20

                                        90-46145
                                          CIP

Preface

The idea of a textbook on design for manufacture arose from discussions centred at the Open University where several academics (from the OU and the Cranfield Institute of Technology) had got together to look at the feasibility of, and requirements for, producing a distance learning course on this subject. At that time, apart from the general climate appearing right for such a course, visits to manufacturing industry to gauge the response, and to consult on requirements, confirmed that education and training was urgently needed on this rapidly emerging and important subject.

Although many industrialists – practising designers, engineering managers, and senior managers – were openly discussing the importance of design for manufacture, with a small number actually applying it, and researchers were active in developing methodologies, techniques and tools for applying design for manufacture, little was available in the way of design for manufacture methods, educational or training courses, or just textbooks that provided adequate coverage of the developing subject.

In Open University terms the textbook was to be designed to act as a Reader to support the main teaching texts of a design for manufacture course. Despite such a course not being available at the present time, the editors took the step now of putting together this textbook as a worthwhile contribution to design for manufacture in its own right. The book is intended to promote an improved and a more structured understanding of the inter-relationships between design and manufacture. As you will find when you get into the book, design for manufacture exists at different levels. To be implemented thoroughly in a company it also has to be applied at these different levels. Applying any one component in the full complement of strategies, principles and techniques is not adequate. To be effective, design for manufacture requires a concerted approach. Towards this end the book makes its own contribution by pulling together a number of written articles from experts and authorities in the field. It contains case studies (a major one from GEC Avery), describes the principles, methods and techniques that are the more formal aspects to design for manufacture and discusses management strategies that provide a framework for positioning all the various components.

Thus in effect the book makes an educational contribution, by delivering experiences and knowledge to engineering designers and students of manufacturing method through the medium of the written word. The articles describe state-of-the-art developments in design for manufacture for the engineering industry at all levels, with Part 4 (computer-aided design for manufacture techniques) describing in particular the applications of knowledge engineering and expert systems. Breadth of coverage is the book's main strength, we believe. As editors we have deliberately included individual articles that collectively represent the spectrum of modern thought on design for manufacture, and which when combined into a volume of contributions, will act as a valuable reference text.

In conclusion the editors want to acknowledge a number of very helpful people who have played their part in putting together this book:

- Isabelle Bricaud, Caroline Davies, Pat Dendy and Carole Marshall who successively provided secretarial support at the Open University;
- Denise Rose who provided secretarial support at Cranfield Institute of Technology;
- Stephen Bishop, Sheila Chatten and Tim Pitts of Addison-Wesley who provided advice and support from our publisher;
- the original authors and publishers who gave permission for their work to be published;
- the staff of GEC Avery who cooperated in the provision of data information and illustrations;
- and finally the Open University and Cranfield Institute of Technology who actively encouraged us to develop this collaborative project.

John Corbett
Mike Dooner
John Meleka
Christopher Pym

October, 1990

Contents

A note on language
For reasons of simplicity, the pronoun 'he' is used to relate to both male and female in this book.

Part 1

Introduction to DFM Based on a Study of GEC Avery's A600 Retail Counter Scale

Designers the world over are primarily concerned with meeting the functional and performance objectives of the product of their design. And rightly so, because if they did not, who would? If that was all that was involved, the designer would have an easy and even a pleasurable task. In fact, some designers of advanced research products do enjoy such a happy and unconstrained life. But once a product is conceived to be sold to the user public in an open commercial market, the picture changes dramatically and the designer has to perform his task under a number of commercial, operational and other constraints.

Perhaps the severest of such constraints is that of costs, not only manufacturing costs, but also development and product support costs, including warranty. Then there is the question of timing; the product has to be available in the market place at the right time to secure its targeted market share. Clearly, a further severe constraint is available technology, both product technology, including materials, and manufacturing technology. The question of volume and rate of production needs to figure highly in the considerations of the designer. Volume, after all, is a deciding factor in the method of manufacture; a product designed for small-volume production will be quite different from the same one produced on a mass-production scale. The former situation would not justify investment in costly plant and tools, but favours simpler manufacturing procedures, perhaps with a more labour-intensive content. A basic illustration of this point is the use of 'machine-from-solid' procedures for small volumes, as contrasted with forging or die casting for high volumes. There are also such major considerations as product developability over the projected life cycle of the product, and this needs to be considered in parallel with anticipated developments in materials and manufacturing processes.

These factors, and there are more, add up to a tall order. How can a designer, or indeed a design team, be expected to be alive in adequate measure to all of these often conflicting requirements? In what priority listing should they be considered? And is such a list a static one? These truly central questions will be considered in a number of papers presented in this book, and a few alternative solutions will be put forward. But arguably, the most significant of all of these factors is manufacturing. There is no escape from the fact that a product designed with manufacturability in mind is more likely to succeed in the market place from *both* a commercial point of view *and* a product performance point of view, including the key issue of product dependability.

It could be rightly argued that *design for manufacturability*, or for manufacture in short, DFM, is such an obvious objective that any designer worth his salt would do so as a matter of course. Why, then, is there such a surge of interest in the subject? It needs to be pointed out that over the last five years or so, a few hundred papers have been published dealing with various aspects of DFM. In fact, the papers presented in this book were creamed off the vast published volume of work. But why the sudden interest?

In the calm days up to the 1960s, international competition was relatively mild and certain economic inefficiencies could be tolerated. The manufacturing scene, both from the technological and operational points of view, was quite simple. The product designer was probably justified then in assuming that the 'production boys' would cope as they had always done. But things are different now; new products requiring new and demanding manufacturing technologies are emerging almost every day. The scale of human activity has multiplied many fold, bringing with it enormous business opportunities. On the manufacturing front, the development of intelligent automation and the use of computers to optimize production also offer hitherto unknown levels of efficiency. The buying public has come to expect near perfect product dependability. Stringent product liability laws are in operation in the USA and are being drafted by the European Community. And then there is the ever accelerating rate of introducing new products or derivatives of existing ones, which serve the business objectives of the manufacturers and respond to the aspirations, defined or not so clearly defined, by the buying public. It is reported that a watch manufacturer in Japan introduces a new model at the rate of one every eight hours! All this is a far cry from the old ways of design for functionality and performance. And unless the product is designed for manufacture, the above objectives are clearly unattainable.

Enough has been said so far in a general context. We need now to secure a sharper picture and bring the issues into focus. The Editors felt that this could be conveniently achieved by presenting a case study that could provide a framework from which to introduce the main issues of DFM. Of course, no case study, however carefully selected, can contain *all* aspects of the subject. The one provided in the next few pages will, it is hoped, cover the domain sufficiently widely to provide a useful introduction. The main body of the book will take the subject further and deeper.

■ Product and manufacturing evolution of weighing devices

From time immemorial, products and produce have been weighed utilizing the comparator principle, by means of mechanical arrangements incorporating levers and knife edges. The desire for weight and price display in the retail business was met in the 1960s by the imposing fan scale (Figure 1). This evolved soon after to the optical analogue display systems (Figure 2), with both weight and price displayed to both customer and vendor. These mechanical machines were labour intensive to produce, requiring high manual skills in their manufacture. The optical system required the application of photoetching techniques for the processing of graticules.

In the early 1970s, electronic display systems were introduced based

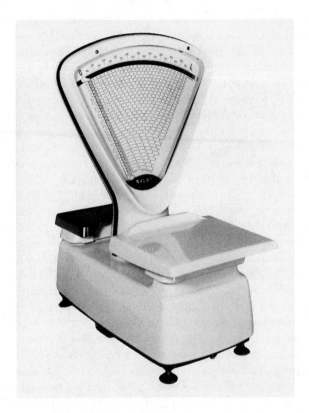

Figure 1 Price-indicating lever mechanism counter scale with specimen chart.

on the flexure of load-sensitive elements in the weighing operation, in place of the use of levers. No computing was carried out; this had to await the arrival of the microprocessor at the sufficiently low prices demanded by the retail counter customer. This significant development coincided with the introduction of strain gauge load cells as the load-measuring element providing electrical signal output in proportion to the weight placed on the scales. Now this represented dramatic changes in both design and manufacturing technologies; the product had moved from being basically mechanical, utilizing lever mechanics and optical measurement, to a mechatronic product utilizing strain gauge force measurement and electronic data processing.

These new products presented their own challenges, but there were great new opportunities as well. Dealing with the opportunities first, it is now possible to *design for function* such that both the electronic data processing and display features are produced to meet specific market requirements. It is worth noting that to cover the needs of the diverse world markets for retail scales in terms of weighing capacity, accuracy, weight increments,

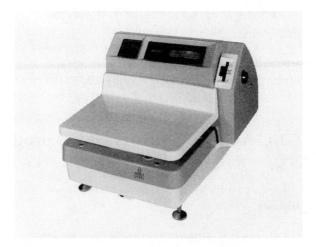

Figure 2 Optical projection indicator counter scale with viewing screen.

keyboards and displays, one product will be produced in 50 different models. This product flexibility has become feasible, both technically and economically, by utilizing the inherent flexibility of computer-aided design, CAD, and intelligent automation in manufacturing. For example, the Avery A600 is produced with over 100 types of screen and keyboard overlays, designed and developed by a CAD system that is directly linked to a photoplotter located in the screen–printing operation. This direct coupling of the design and manufacturing operations has provided benefits well beyond the original objective of product variety in that both design turnaround times and manufacturing lead times have been reduced from weeks to days. There are also improvements to product quality as compared with draughtsmen-produced designs. A similar story can be told regarding the alternatives of electronic computing and the manufacture of mechanical components.

   As to the challenges of these new technologies, they can be broadly divided into human and investment challenges. The industry had to develop or acquire the know-how and skills relating to electronic element design and manufacture, as well as recent advances in mechanical components, their materials and manufacturing processes. In addition, the very organizational structure of a company in this industry had to change to reflect these new technologies. The implications of these changes on all employees were bound to be profound.

   As for investment, new technologies demand new design and manufacturing methods which, in turn, require the acquisition of new, often costly, plant and machines. And all such investment in people's skills and in capital has to be provided within an economic framework such that the superior product can be made available to the buying public at significantly lower prices. And this is where DFM comes into force. Since, if the designer gets carried away with the new technologies and the opportunities they can undoubtedly provide, manufacturing costs can very readily escalate. It is probably appropriate at this juncture to move on from generalities to the specific situations encountered within the company itself.

■ The GEC Avery Company and its product range

☐ Recent history

W&T Avery Ltd is the largest British weighing machine manufacturer and has held such a leading position over a period of more than 200 years. The company currently employs over 2500 people in the UK, with operations throughout the world. A further 3500 people are employed in 15 overseas subsidiaries. Avery supplies some 70 000 units per annum, which represents some 20% share of the total world market.

The leading position the company enjoyed for a long time had its foundation in the design and manufacturing skills relating to mechanical weighing systems. In 1972, Avery produced the world's first electronic scales, for which the company received the Design Council's award in 1974. Yet in 1979, when Avery was acquired by the General Electric Company, GEC, 90% of Avery's products were still mechanical, with skills and a manufacturing base geared to this type of product. This was at a time when the demand for mechanical retail scales was declining at a rate of more than 20% per annum. By that time, it was clear that the somewhat slow growth of the electronic scales sector experienced in the early to mid-1970s would explode by the end of the decade. Thus, the GEC acquisition of Avery coincided with a major technological and commercial change. This change required a corresponding restructuring of the company's management, its operations and investment plans. It is worth noting that competitors were implementing microprocessor technology and modern mass-production techniques which resulted in a steady reduction in product prices by more than 15% per annum. The Japanese had also entered the electronic price-computing retail scale market with machines based on calculator/cash register technology. It was recognized, at that time, that Avery was innovating too slowly to meet these real threats to its existence, and that urgent and radical changes were required to ensure the company's short-term survival and long-term viability.

Following the GEC takeover, a review of the business of the company was undertaken. Central to this review was the recognition that the greatest need was for the introduction of mechatronic products. A small marketing team was created to gather competitors' market data as a starting point for organizing the company's first business and product planning conference. This took place in late 1981.

As will be seen later, the conference, together with the follow-up actions, proved to be truly successful in transferring a company under real threat from its traditional mechanical products to a mechatronics-based company with growing sales and profits. A major factor in the success of the conference, which has since become an annual event and the prime mover in developing business and product strategies, was the involvement of line managers at the operating level from *all* areas of the business. These included home and export sales, product engineering, manufacturing engineering, production management, data processing, finance and service.

An analysis was undertaken whereby the company's business was broken down into individual segments. Key business segments were identified in terms of sales value. From this, 20% of the segments that combined to account for 80% of sales were selected for detailed analysis. Strategic objectives were identified and plans were prepared with recommendations for new product development and market entry programmes. A top management team reviewed all of the recommendations made by the conference, consolidating them into a new product development programme for the company as a whole. This included detailed new product specifications, target selling prices, volumes and manufacturing costs for all key segments of the company's business.

Product plans, however grand, are of little worth unless they are implemented. It is necessary to organize and manage company resources to implement the plan quickly and efficiently to meet defined product timing and cost targets. It was quickly discovered that the existing vertical organization and product development procedures were inappropriate for this to happen. For example, quite separate programmes were generated for home and export products. Sales provided new product specifications to the product design function. When the designs were complete, these were handed over to manufacturing engineering. In fact, there was no liaison between the product design function and manufacturing, merely an 'over-to-you' event at the very end of the design phase. Where was DFM to be found you may well ask? No wonder the development cycle lasted between two and three years and the company then produced on average only one new product every year. These deficiencies were rectified by undertaking three major changes; one organizational the other two operational:

(1) Reorganizing the engineering department: The then existing structure was based on division by discipline – mechanical, electrical, detailed design, and so on. The engineering department was reorganized into four self-contained groups reflecting the primary market sectors – that is, retail, industrial, precision balances and force-sensing transducers. This new arrangement provided product objectivity to the activities of

the engineering department, perhaps at a slightly lower level of specialist staff time utilization.

(2)   Introducing project management teams: Project teams were introduced to ensure that each design team engineered its selected product to the optimum production cost. A key member of the project management team was a manufacturing engineer whose primary role was to see that the product was designed with manufacturing in mind, in other words DFM. This is a far cry from the earlier 'over-to-you' practice! Another key member of the team was an export representative who ensured that the products designed could, from the outset, and with the minimum of hardware modification, meet the weights and measures legislations and the varying market needs of a wide range of countries.

(3)   Implementing new product development programmes: This was achieved by setting up an executive committee of key managers from sales, marketing, engineering and production. The authority of this body was far-reaching, covering the definition of product programme priorities, approval of product development funding and appraisal of progress, as well as the integration of new product ideas into the development process. A new product development procedure was laid down and widely broadcast in the company to ensure that the procedure was understood and followed by all concerned.

Within nine months of establishing the new order, the company began to introduce new products at the rate of two per month. This is to be compared with the previous rate of one per year. The importance of the introduction of new products to the success of the business cannot be over-emphasized, as it is a crucial factor in increasing the competitiveness of the enterprise, and its success invariably leads to an increase in market share. But a high rate of introduction of new products cannot be achieved unless design and manufacture move forward hand in hand. In fact, the company introduced not only the principles of DFM in its new structure, but also the broader concept of *simultaneous engineering*. This concept endeavours to ensure that a new product is engineered so that it meets all of its specified technical and business objectives, including its targeted market share at the appropriate quality level and specified costs. Simultaneous (sometimes referred to as *concurrent*) engineering can be achieved by involving members of functions other than the product design function in the process of design. Such members are sometimes called *co-designers*. DFM is a key element of simultaneous engineering as it represents the co-ordinated activity between product design and manufacturing engineering; this is the axis around which the majority of business objectives revolve.

It is clear from the above that the implementation of the concept of DFM requires major changes in conventional company organization, leading to the creation of multi-disciplinary teams at both planning and operational levels, with appropriate authority vested in them and the full backing of top management.

☐ The product range

As mentioned earlier, the company's operation can be broadly divided into four divisions:

(1)   Retail and retail systems division.

(2)   Industrial and industrial systems division.

(3)   Precision weighing division.

(4)   Force-sensing transducers

The last of these acts as a vendor to the three operating divisions. Before the reorganization, Avery 'bought in' these devices which somewhat restricted its freedom of product development. The presence of the word 'systems' in the titles of the first two divisions hides an enormous technological and business activity, as evidenced by the presence at the point of sale in supermarkets of electronic price-computing scales that link up with central computers and act as truly comprehensive management information systems.

But returning to the business segments themselves, we shall concentrate now on retail scales, for two reasons. Firstly, this product range accounts for the largest single business segment, perhaps a little less than 50%, of the total sales and profits of the company. Secondly, the product selected for this case study, namely the A600, falls within this division; in fact, at the time of writing it is the most recent product to be produced by the company utilizing DFM methods.

To appreciate the nature and magnitude of the novel features introduced in the A600, it is useful to look back into the ancestry of this product.

Brief reference has been made to the earlier mechanical products that evolved into the price-computing 'fan' model, followed by the optical projection model. Then came the 'electronic jump'. The early limited volume requirements for electronic systems were primarily satisfied by outsourcing. But then the company began to experience quality, price and delivery problems. To relieve such problems, and drawing on the in-house expertise gained from photo-etching experience in processing graticules for optical machines, a pilot low-cost processing plant was established for the manufacture of single-sided printed circuit boards (pcbs) using manual track layout and processing techniques. As the market grew,

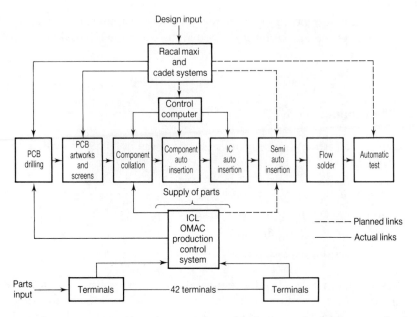

Figure 3   Interrelationships of computers in the design and manufacture of pcbs.

major investments were made in providing automated processing plant for double-sided through hole-plated bare board manufacture, automated assembly of electronic circuits, mass soldering equipment and automated test facilities. Such automated manufacturing facilities have now been integrated with pcb design by a CAD/CAM system, with the whole system being controlled by a computer network, as shown diagrammatically in Figure 3. The company currently produces 8000 boards per week, all designed and manufactured in-house.

The current opportunities in electronics and data processing can serve the needs of the retail market in new ways. Such needs are wide and diverse, ranging from the complex requirements of the multi-supermarket chains incorporating fully networked scales, bar code readers and price-computing printers, all linked to a central computer and regional warehouses. But there is also a vast market required by the independent retailers, such as the corner store or the mobile shop, for a stand-alone, low-cost product. The company developed a product range, designated the 1700 series, culminating in the 1770 model, termed the 'Independent', which was introduced in 1983 (Figure 4). This is the model that can be seen currently in many shops in the UK and throughout the world.

It was clear soon after this development that world market volumes would increase dramatically, so the company decided to secure an increasing share in that potentially lucrative market. It was equally clear that competi-

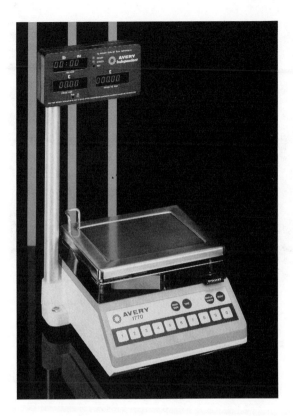

Figure 4   The 'Independent' model 1770.

tion was likely to be severe and that the market would be particularly price sensitive. A new product range was conceived in 1986, designated the A600, which was to be designed with manufacturing cost very much in mind.

The first such product was the A600 itself (Figure 5), which will be contrasted with the functionally compatible 1770 model, to reveal how GEC Avery applied DFM principles to achieve its business objectives.

■ How the A600 was designed for manufacture

Here we aim to follow the progression of design of the product at the crucial concept stage, thus illustrating one approach used to apply the principles of DFM. We shall then proceed to examine the outcome of such design for the A600 and make comparisons with the 1770. But first, we need to look briefly at the general construction of a typical product.

Figure 5   The A600, designed for manufacture.

☐   General construction

A typical price-computing and weighing device has the following basic construction, as shown diagrammatically in Figure 6:

- a rigid base;
- the load cell, which is rigidly fixed to the base;
- a 'cross' – this is the intermediate component resting on the load cell and supporting the pan above;
- a cover to protect the transducer and electronics;
- the pan, which contains the product to be weighed – it can be a flat plate, a scoop or an oval pan, all of which are generally made of stainless steel.

In addition, there is a membrane keyboard and two displays – one for the vendor, the other for the customer. The main body of the scales also houses the various electronic boards and electrical connectors.

☐   The design concept stage

The design team had their main objectives clearly defined for this new product; it had to be designed for minimum manufacturing cost, without

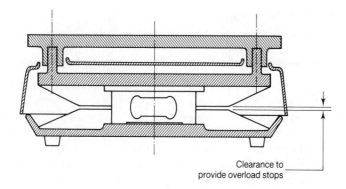

Clearance to
provide overload stops

Figure 6 General construction of a counter scale.

compromising funtionality, quality or adherence to weights and measures laws. The route to achieving these objectives was to be DFM. But the question was: Where to begin?

One reasonable approach might be to examine the individual components of the existing product, namely the 1770, and to endeavour to reduce their costs of manufacture. This approach may seem to be appropriate, but it has the inherent flaw that when the modified constituent elements are brought together, certain serious mismatches might emerge and a long cycle of iterations might then be required, with no assurance of ultimate solution. Rather, the decision was taken to pursue a top-down approach and to view the product as an entity, concentrating primarily on its final assembly. It is possible to quantify the 'efficiency' of a design in terms of final assembly time. This is of crucial cost importance to the original equipment manufacturer, OEM, who generally assembles the final product from components supplied by vendors or in-house. Basically, a reduction in assembly time and cost can be most effectively achieved by reducing the number of parts that go into the completed product. Consequently, the number of fasteners is also reduced.

But the very fact that the constituent elements of the product are being examined with part integration in mind demands a 'rethink' of the design, materials and method of manufacture of the new integrated parts. The outcome therefore is a procedure that aims to reduce the time and cost of assembly and to assess new approaches to part integration while at the same time ensuring that the final product will fit together nicely.

Let us now examine the application of this methodology to certain aspects of the concept stage of the design of the XA1, which is the pre-designation of the A600.

Basic assembly options

Three options were examined: housing build, upside–down build and base build (Figure 7). Note that in each case the number of major components is identified together with the number of screws and studs. The arrangement selected was the base build, which allows 'layered' assembly procedure, feeding components to the base in one direction, namely vertically down.

At this key stage, the designer is already taking wider issues into account – for example, material selection for key components and the feasibility of automated or robotic assembly. The designer is anticipating high-volume production when the cost of automated assembly plant would be justified by the high-volume output. But even if such a stage is not reached, a product designed for automated assembly would be easier, and hence less costly and more reliably assembled manually.

The scale base

In the 1770 and indeed all predecessors, the base was constructed in an aluminium casting, or steel fabrication. The designer was, however, aware of developments in polymer engineering and the availability of rigid thermo-plastic mouldings – for example, in the manufacture of garden furniture and other applications. Rigidity of the base is essential, since deflections could generate slight displacements in the transducer, leading to inaccuracies in weight readings. Stiffness can be provided by means of stiffening ribs incorporated in the moulding.

The material finally selected was a high-impact strength polystyrene utilizing the 'cinpres' injection-moulding process, which is capable of producing low internal stress mouldings and constant wall thickness with hollow stiffening ribs. The configuration of the base moulding is shown in Figure 8. This configuration reduced component and fastening count by 30 and also the cost of purchasing and stock control of the numerous 'bits' that go into the manufacture of an equivalent metallic product. Further, the material cost of a moulding is lower, and machining and finishing operations are eliminated.

There were, however, two difficulties, one specific to this product, the other of a more general nature. The electrically conductive metallic bases are capable of screening the electronic elements against radio frequency, RF, disturbances. It was therefore necessary to spray the polymer moulding with a thin metallic material to secure similar protection.

The second more general difficulty of part integration by means of the manufacture of a complex injection moulding is that modification to the design can prove to be quite costly. This is because the injection-moulding tools have to be redesigned and new ones produced. Therefore, if the injection-moulding course is to be followed, the design must be truly firm. Further, it is prudent to introduce some features that seem to be redundant initially, but may be required for a later version of the product.

(a)

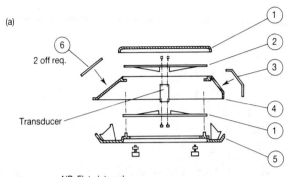

6
2 off req.

Transducer

1
2
3
4
1
5

NB Flat plate only
Major component count: 8 + keyboard

(b)

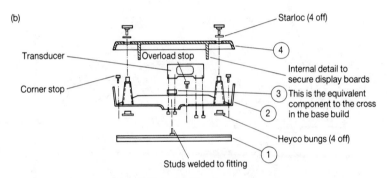

Starloc (4 off)

Transducer
Overload stop

Corner stop

4

Internal detail to
secure display boards

3 This is the equivalent
component to the cross
2 in the base build

Heyco bungs (4 off)

1

Studs welded to fitting

+ 2 screens integral with keyboard overlay

Housing and base (2 and 4) held by starloc fasteners
4 screw (assembly) + 4 corner stops and 1 centre stop
Major component count: 6 + keyboard

(c)

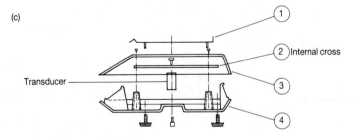

1
2 Internal cross

Transducer

3

4

NB Screens integral with keyboard

Figure 7 Basic assembly concepts: (a) housing build; (b) upside-down build;
(c) base build.

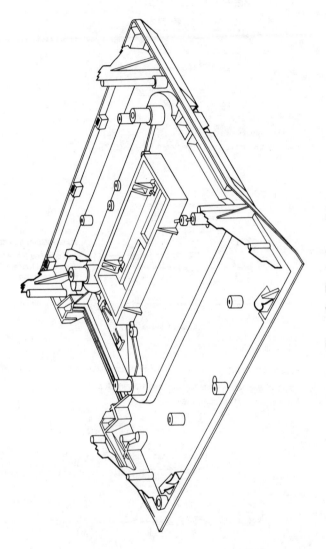

Figure 8 The base moulding.

It is also worth noting that polymer components can incorporate 'snap-fit' features, which can assist considerably in the assembly operation and reduce the need for conventional fasteners such as screws.

The viewing angle

The viewing angle is an interesting ergonomic consideration that may not at first appear to have DFM implications. The USA prefers to have its counter scales at a higher level than in Europe and the rest of the world. The inclination of the display boards for both vendor and customer are at a larger angle to the horizontal in the USA than in the rest of the world. On the vendor's side, the corresponding angles are 67° for the USA and 20° for Europe (Figure 9). Avery had produced their scales to these requirements for many years.

An ergonomic study was undertaken in an endeavour to rationalize this situation. It was concluded that an angle of 35° was acceptable to both markets and the A600 was so designed. This is a simple yet significant DFM decision with beneficial repercussions to manufacturing in both component design and assembly, as well as the purchasing function and stock control.

Transducer mounting

The transducer has to be securely mounted to the rigid base to ensure accurate weighing. In the 1770, the transducer was bolted to the metallic base. The tightening torque of the bolts was measured to ensure adequate mounting without undue distortion. This was a slow and laborious operation.

Various alternatives were considered (Figure 10). The one selected was a drop-and-slide arrangement to secure an interference fit under moulded lips in the scale base. Only one screw was used to secure the mounting.

Scale services bracket

Although it was originally conceived to integrate mains supply input and battery housing only, the decision was taken to design a *universal* bracket to provide flexibility to configure the following options, in any combination:

- mains cable supply;
- mains switch;
- mains fuse;
- battery supply socket;
- battery supply switch;
- printer interface socket.

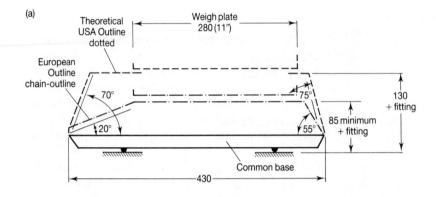

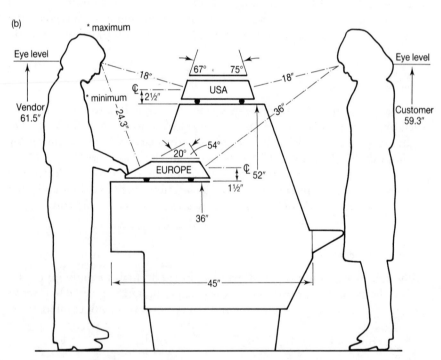

Figure 9   The viewing angle: (a) product geometry; (b) operational arrangement.

(a)

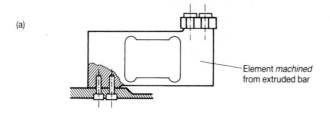

Element *machined* from extruded bar

(b)

Tee slots extruded

Moulded end brackets

Nut plate

Fixing screw

Single screw fixing No machining

NB High clamping force *will* require plate to 'spread the load'

– Robotic assembly ?

Extruded element

– No drilling of element

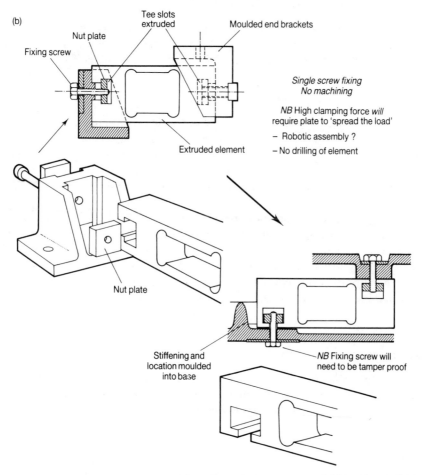

Nut plate

Stiffening and location moulded into base

NB Fixing screw will need to be tamper proof

Figure 10 Transducer mounting: (a) traditional arrangement; (b) 'drop-and-slide' arrangement.

A *single*-moulded bracket incorporating *all* of the apertures to accommodate these features was designed, utilizing adhesive-backed blanking plates to seal off apertures not required by the customer. The significant cost and operational advantages are obvious.

Although it was possible to include the bracket as an integral feature of the scale base moulding, it was felt that this would cause significant complication to the injection-moulding tool, necessitating the introduction of moving cores. Further, the services bracket can be more readily incorporated in the customers' specific electrical supply option.

Mounting of electronic and display boards

In the 1770, these boards were mounted to the base unit by means of screws. These have been eliminated in the A600 by the provision of clip-in features in the moulded base unit. Assembly of these boards is completed in seconds.

Switchless electronic configuring and calibration

The configuring of the scales to satisfy the varied requirements of the more than 30 countries can be a complex and laborious operation. These requirements include local weights and measures regulations, currency and language. In the 1770 and its predecessors, this was achieved by operating internal mini-switches in a certain sequence and adjusting wiring links. In the A600, the switches have been eliminated with obvious cost and assembly advantages. The distributors or operators can also benefit in that to secure the desired configuration, utilizing built-in software, it is now necessary simply to press keys in the specified combination. It is now planned to so configure scales on the production line by linking a small computer to the assembly line. A calibration procedure is incorporated with the configuring cycle, and it is all done automatically. An exploded view of the A600 is shown in Figure 11.

Summary

The steps outlined had profound effects in reducing manufacturing and assembly costs, as can be seen in the following section. A change in attitude has now been established and every design is evaluated at the concept stage by defining:

- a parts count monitor;
- a design for assembly rating, namely assembly time.

The combination of these two factors is an important measure of successful DFM. But in the final analysis, it is manufacturing costs that will have to be calculated, taking all relevant factors into account.

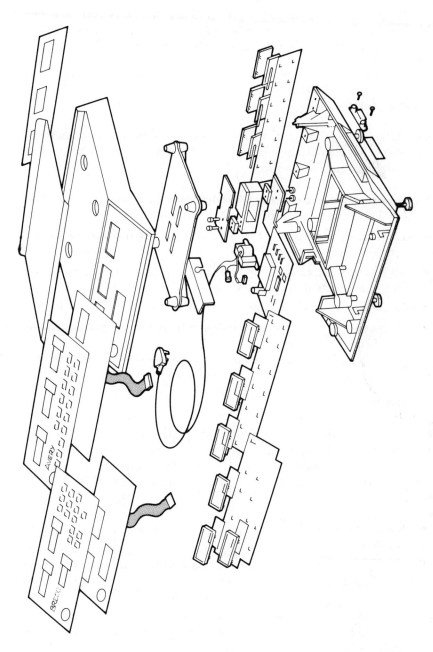

Figure 11 Exploded view of the A600.

■   The benefits of DFM to the A600

There have been significant and, in certain instances, spectacular benefits, justifying the efforts of those actively concerned. We shall briefly examine these benefits by comparing the A600 with the 1770.

☐   Parts count

It took 110 parts to complete the 1770 including 62 fasteners. For the A600, these figures were reduced to 52 and 18 only, respectively. But there are considerable hidden benefits also:

- Fewer bought–in parts reduce the operational burden on the purchasing department with corresponding cost reduction.
- Bank charges are reduced as a result of fewer transactions by the purchasing department.
- Inventory is also reduced, resulting in a reduction in storage space and part transport within the plant.
- Production control's task is much eased, as is the risk of late delivery.
- With fewer parts, the risk of deviation is proportionately reduced, thus enhancing quality assurance.

☐   Assembly time

The assembly time for the 1770 stabilized at 0.7 hours. For the A600, this was reduced to 0.3 hours. Figure 12 shows the reduction in assembly time for a range of models, including the 1770 and the A600. Clearly, there are direct labour cost benefits. But again, there are also a number of hidden, potential benefits:

- Space: With less assembly work to do, the assembly task can be conducted in a smaller work area.
- Assembly equipment: Even manual assembly operations require some machine aids. With fewer assembly operations, the cost of such machine aids will be correspondingly reduced.
- Possible assembly automation: It will be easier to justify capital investment in automated assembly with fewer assembly tasks to be undertaken.

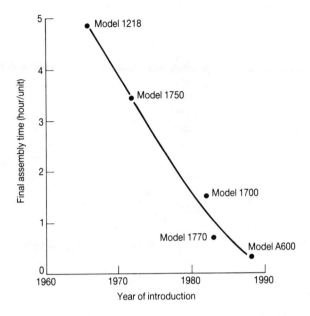

Figure 12   Reduction in final assembly time for a range of models.

☐ Product weight

A spectacular weight reduction was achieved from 10 kg for the 1770 to 4.8 kg for the A600. True, this was primarily due to the use of polymers, but again they were introduced as a DFM objective. Consequently, significant reductions in the cost of packaging and shipping, also damage in transit, were experienced.

☐ Manufacturing costs

It is not permissible to divulge manufacturing costs in an open publication; the most that can be revealed is the selling price. Figure 13 shows the significant reduction in selling price for the three models 1700, 1770 and the A600. Such a price reduction will undoubtedly continue, and it is also reasonable to expect, in parallel, an increase in product performance and functionality.

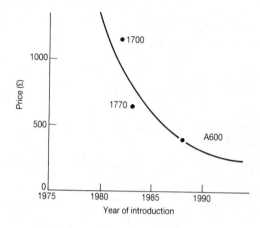

Figure 13 Reduction in selling price for a range of models.

■ Concluding remarks

The Avery case study demonstrates that products can indeed be designed for manufacture. Numerous business benefits can be derived from so doing. Yet DFM cannot operate in isolation; it needs to be conceived and implemented as part of a broad strategy including sound product planning and operational management. Leadership and commitment from the top is clearly essential. All involved need to recognize that design is too important to be left to designers alone. All designers and manufacturing engineers need to be trained in DFM.

The forward move of a product such as the 1770 and the A600 is influenced by two sets of forces: technology push and market pull. Figure 14 illustrates these forces for the electronic retail scale market. What DFM does is to translate market pull factors into the push of enabling technologies for both product and manufacturing alike.

■ Reviewing Avery's needs

The above case study was provided to set the scene for DFM by describing the relevant experiences and achievements of a single company. The approaches used by GEC Avery were almost exclusively self-generated from within. Had a published DFM body of knowledge existed at the time, Avery would have probably gone further in their endeavours, or indeed followed different lines of approach. But a considerable amount of published literature

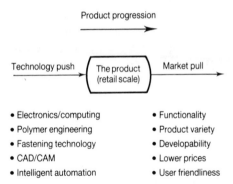

Figure 14   The push–pull forces of product developments.

is now available and it would be useful to view Avery's needs in implementing DFM at the time with the accumulated knowledge of the subject now. Such an exercise would also form a convenient bridge between the case study and the main body of this book.

The majority of papers presented in this book emphasize the crucial importance of two organizational and managerial issues: firstly, the setting up of multi-disciplinary teams and, secondly, the positive involvement and commitment of top management. Well, Avery adequately attended to both issues and this has developed into a day-to-day way of life. But there is more to it than that. Avery would have benefited from further thoughts and experiences regarding the operations of an interdisciplinary team. For example, how to get independent-minded specialists to work closely together? Who is likely to be the most effective team leader? What is the role of the strategic planning department? Regarding resource allocation by the team to the operating departments, should more resources be provided in the early design phase? These and related issues are dealt with in Parts 2 and 5 of the book.

There are currently a number of quantitative methods that can be of material help in implementing DFM. Avery effectively used the Boothroyd and Dewhurst method which focuses on final assembly and also quantifies component assembly features and their costs. But there are other methods. There are also related methods that assist in securing robust design – for example, the Taguchi method for statistically planning experiments to reveal the most significant design parameters. There is also the quality function deployment, QFD, method which refers design decisions to the requirements of the customer, whether it is the ultimate user customer or an internal 'customer' within the company. These and other methods are covered in Part 3 of the book.

Finally, computers have come to the aid of DFM in two significant ways: firstly, as a means of constructing databases and, secondly, as interactive knowledge bases. Both are useful in bringing information and knowledge to the attention of both designer and manufacturing engineer. The availability of both computer systems and software is increasing rapidly and Avery would have benefited from these opportunities, had they been available at the time. Part 4 presents the state of the art of computer-aided DFM techniques.

Part 2

Principles of DFM

■ Introduction

All of the authors of papers in this part stress the need to form multi-disciplinary design teams at the start of the design process. This is in contrast to the step-by-step approach practised in the past by most manufacturing companies in which the design passed from design to process planning, to manufacturing and then to sales, in a sequential manner. A major recommendation is for a product to be designed in parallel with the design of the manufacturing system, or alternatively to be compatible with the existing manufacturing facilities. This *simultaneous* approach obviously makes sense when considering the point made by Whitney that between 70 and 80% of the final production costs may be determined by design. The formation of multi-disciplinary teams is an important factor in improving communications and considering various concepts simultaneously. Integrated and balanced teams are required to address the facts that:

- the product has to work;
- it has to be manufactured;
- it has to sell;
- it must be profitable.

In the past, the design of products has usually been completed before manufacturing is considered. Attempts to introduce value engineering techniques generally concentrated on reducing costs after the design was finished. In contrast, the papers presented in this part emphasized the need for cost effectiveness to be achieved during the conceptual stage, and then to continue throughout the design, development, production and complete life cycle of the product. A major problem with value engineering was the way in which it was implemented into companies rather than in the technique itself, which, if properly applied, encompasses all of the principles dealt with in this part of the book. Value engineering tended to be placed outside the mainstream company activities (Figure 1) and products were able to proceed directly from design to manufacture, bypassing the value engineering section. Further problems arose from the fact that value engineering was used to challenge direct material and labour costs rather than the design concept, where the most significant savings could have been achieved. Although companies claimed initial success with value engineering programmes, it was often the case that many of the recommendations were not considered sufficiently worth while for implementation, as to do so extended time-scales. Clearly, the worthwhile recommendations should be incorporated before the product goes into production, because modifications are more expensive at this late stage and are most unwelcome once the products are being demanded in production quantities. Modifications also delay the launch of a new product with the result that the product is late on to the

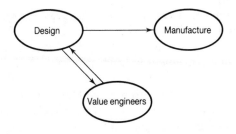

Figure 1   Value engineering in the 1960s.

market. 'After-the-fact' engineering increases costs through the duplication of engineering effort, which would not be necessary if the job was done right first time.

■ Discussion of papers

The first paper in this part, by Whitney, covers many aspects of DFM, from organizational strategies to fundamental design principles, which are re-emphasized in the other papers in this part – for example, designing for modular assembly, minimum parts count and flexible manufacturing methods. The trend in manufacturing companies to reduce inventory costs, through producing smaller batches, has resulted in a need to minimize tooling and set-up times. Clearly, in order to maximize the potential benefits of these modern manufacturing trends, the design of the parts is critical. A particularly good example of designing parts for jigless manufacture is included. This technique minimized tooling set-up times and enabled the company to respond to dynamic markets, which required small batches and short lead times, without carrying expensive stocks. In addition, the paper emphasizes the need to design quality into the product, rather than inspect it in later.

The pay-offs of preventing problems during the design stage are, unfortunately, difficult to quantify. Nevertheless, although the actual values may be subject to debate, there can be little doubt that it makes good economic sense to move problems further upstream. Unfortunately, many companies find that this is a difficult concept to adopt and they continue to 'fire fight' in production. This is understandable, although no excuse, when one considers the difficulty of knowing what problems were actually prevented upstream. The inevitable result is that designs, which were assumed to be complete when leaving the design department, require changes, which consume valuable resources. Effective communications between design and production engineers, for example, mean that the manufacture of the product is considered as soon as design commences.

Involving production engineers earlier has the advantage that it gives them a sense of ownership in the design of the product.

Whitney also discusses design as a strategic activity and the necessity for top management involvement and commitment in order to implement an effective DFM strategy. Without this involvement, conflicting traditional departmental objectives are likely to persist. Therefore, an important step is to improve communications between all team members and encourage the use of less formal systems, to ensure that they do not feel threatened through sharing their own specialist expertise.

Reasons for products failing in the market place include:

- poor quality;
- poor timing;
- poor communications;
- unexpected competition;
- inadequate investment;
- high product cost.

Senior management is ultimately responsible for these factors, and although the blame is often given as inadequate design, it is in reality poor management of the design process.

It is no use merely stating objectives and then assuming they will automatically be carried out. They will be met only as a result of a conscious and determined effort. One of the first objectives should be to prepare a product design specification that is detailed and clearly written, in order to avoid any ambiguity. Unfortunately, many companies place too little emphasis on compiling a well-defined specification before design commences. While a clear, well-thought-out specification does not guarantee high-quality competitive products, without such a document the chances of success are severely limited. The specification should not dictate the design, but it should discipline the designer to think in terms of market requirements. Therefore, important features of a complete design specification are:

- the ability to meet the reasonable and foreseeable needs of the market;
- to contain information on costs and quantities.

Quantities are important as they have considerable influence on the manufacturing process to be utilized, which in turn affects the design of the product. For example, large numbers may be required to justify investment in special tooling or automation equipment.

In recent years, many companies have introduced automation into the assembly area with varying degrees of success and failure. The paper by Corbett expands on the design for assembly issues presented by Whitney and discusses additional aspects of assembly, including the relative merits of adhesive bonding and mechanical fastenings. He also includes design checklists for manual and automatic assembly, and case studies arising from the use of the 'design for assembly methods', described in Part 3. The checklists include pointers to simplify component assembly while the case studies emphasize the need to reduce the number of parts, in order to eliminate assembly operations and gain the largest cost savings. However, in design for manufacture there are no all-embracing rules. For example, General Motors in the USA (Eaton, 1987) has designed a new transmission case design that utilizes three parts to replace a one-piece casting, thus going against the normal aim of parts reduction for assembly. The design team worked closely with the supplier of the aluminium castings for the case design. The result was a novel design that overcame the usual problems of casting porosity, which requires a remedial impregnation process. In addition, the casting was purchased 'close to finished form', which minimized machining. The result of reducing material costs by over 50%, reducing scrap by over 60% and the number of tools utilized by over 30% vindicated their approach. Whitney also gives an example in which a modular design resulted in assemblies with more parts but with lower manufacturing costs coupled with improved quality. Designs, clearly, ought to be thoroughly reviewed in order to maximize the potential of available manufacturing processes while ensuring the minimum number of parts to achieve the functional requirement at minimum cost.

Martin (1988) questions the need for formal design reviews, indicating that it is important to get the team in at the concept stage and to continue reviews as 'a day-to-day job'. It is true that design reviews can waste a great deal of time if management is weak. This often leads to the over-preparation and rehearsing of presentations, apparently to boost the managers' confidence. Obviously, this is a waste of time and effort and, in such cases, reviews can set a project back several days or even weeks. However, a well-conducted review that is kept under firm and not dictatorial control provides valuable information on the state of the project, against which decisions can be made. In particular, reviews provide:

- a check against the product design specification;
- an indication of the product cost and profit potential;
- an opportunity for the team members to optimize function, cost, reliability and appearance.

In order to gain maximum benefits from the review, the participants should receive relevant information well before the meeting. If there is no time to review the data prior to the meeting, much time will be taken up understanding basic concepts and questioning minor detailed points. This leads to the risk of missing any major potential improvements to the design.

A badly managed review can lose all of its potental benefits. It is important not to take on the appearance of a board of enquiry, with the designer on trial.

The next two papers discuss designing for specific manufacturing methods – CNC machining and advanced castings. In the first of these, Hodgson and Pitts include some sound advice when designing for CNC manufacturing methods. This involves the need to minimize part transfers between machines, the number of tools used and set-ups. Milling and turning examples are included which emphasize the rule 'never deviate from the primary axis' in order to obtain single set-up, which is often referred to as the *one-hit machining* method. This method is used to machine complex multi-faced components on just one machine, thus avoiding the stop–go delays inherent when parts are moved from machine to machine. Overall benefits claimed following introduction of *single set-up machining* are:

- reduced batch sizes, resulting in greater flexibility in responding to fluctuating demand;
- reduced labour, as combining several operations means fewer movements between machines;
- reduced scrap, fewer operations mean fewer errors;
- reduced floor-to-floor times;
- reduced inventory;
- reduced lead times to customer.

Single set-up machining used in combination with measuring probes can dramatically reduce the costs of fixtures, as the probes can readily identify the component's position, eliminating the need for precise location. Figure 2 illustrates an example of a component produced on a four-axis machining centre which is machined on all surfaces. Probing is undertaken initially to find the true centre of rotation of the fourth axis, which can be used as a set-up function. The probes are used later to measure critical dimensions, so that tooling offsets can be updated. Single set-up machining of the components shown in this figure resulted in the reduction of set-ups and machining operations from seven to just one. This was coupled with an 80% reduction in work in progress and a large reduction in scrap.

Hodgson and Pitts discuss the advantages of single set-ups in responding to requirements from sales, which are increasingly calling for a rapid manufacturing response and frequent changes to products, in order to satisfy customers' demands. However, it is essential that the products are designed for the appropriate manufacturing methods from the outset, in order to maximize the advantages offered by these new techniques.

Mietrach's paper relates to aircraft components. However, the principles expounded are sound for other manufacturing sectors. The paper describes the recent advances made in sand casting, investment casting and précial casting, and indicates the relative merits of each process. Recent

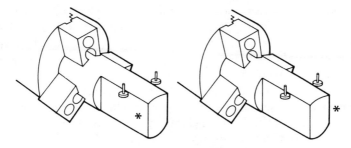

(a) Centre line position determined by averaging

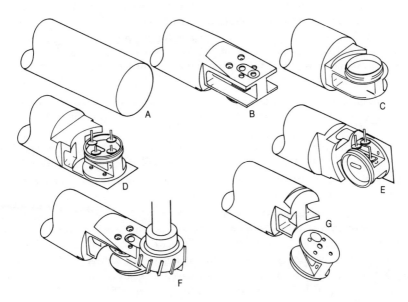

(b) 'One hit' method

Figure 2   Single set-up machining of components.

advances in these processes have resulted in improved accuracies, which means that parts can be cast close to their finished form, which minimizes or eliminates machining.

Useful hints and examples are included. In one of these, a 35% cost saving was achieved with no weight penalty, by using a single casting to replace 400 parts and 2000 fasteners. Another example shows how a cost reduction of more than 60% was achieved by modifying the design for castability. Clearly, in order to achieve this scale of saving, without loss of function, it is important that the designer designs for casting from the outset, and consults the foundry. Mietrach makes the point that the cost of a casting can easily be doubled by factors controlled by the designer.

The last two papers included in this part describe specific case studies, which illustrate different aspects of DFM. In the first of these, Gill shows how many of the principles, discussed in earlier papers, have been used to reduce the work's cost and improve the market share of a range of squirrel-caged induction motors. Before the study, the range included a large number of variants, a large number of dies and an ever growing inventory. By understanding the range, adopting a 'modular' design approach, rationalizing part count and casting close to finished form, the manufacturing costs were reduced and the improved casting accuracy resulted in a weight reduction of 20%, as well as reducing machining. Gill emphasizes that the whole initiative was 'design led, but throughout there was consultation with other departments, particularly marketing and manufacturing'.

There are positive advantages to be gained from having a single leader, or product champion, responsible for the overall project. These include:

- improved co-ordination between team members from various departments;
- decisions being made with the success of the project being the main criteria, rather than the interest of specific departments.

Without such a person, it is difficult to direct and control projects effectively or retain team spirit. The project leader needs to be the main contact with top management, departmental managers and the customer.

In the final paper in this part, Goldstein highlights the need to reduce lead times, in order to get new products on to the market ahead of the competition. It therefore makes sense to allocate more resources during the early design phase of a project, which may incur more initial expense but will be less costly than allocating additional resources to rectify problems later, and thus delay the launch date. It is easier to gain sales by being first into the market than it is to win them back following a late entry. The ultimate aim must be for a company to make its own product obsolete before this is done by the competition.

Goldstein describes the benefits the company has derived from the use of a computer-aided process planning system, which has proven to be extremely effective in providing fabrication cost estimates at an early stage in the design process. The availability of early cost data is extremely valuable for DFM programmes, and is discussed by Boothroyd and Dewhurst in more detail in Part 3. The availability of reliable cost data is essential when designing to cost.

For competitive design, costs need to be considered as important design parameters, which are included in the specification alongside the functional requirements. The outcome of a design-to-cost programme should be a product that meets the specification in providing an adequate performance, rather than the best possible performance, at a cost that will succeed in the market place. It is unacceptable to provide less than an adequate performance, and equally unacceptable to provide additional unspecified features at extra cost. However, if a designer is required to design to cost, he must be supported by accurate cost data and estimates. Successful design–to–cost programmes require close departmental collaboration, in order to obtain satisfactory cost estimates early in the design process. Estimates cannot be made without some form of design input.

The main problem is how to obtain the quality of estimate required early in the conceptual design stage. Inaccurate cost estimates can affect the perceived viability of projects, and also increase the project and final product cost. For example, overestimates can result in allocating excessive resources to a project, while underestimates lead to under-resourcing, which causes reorganization of the project, which is expensive in terms of both time and money. Improved cost estimates can be obtained if estimating is taken seriously and all causes of estimating errors are investigated. It is important not to apply pressure in an attempt to reduce estimates through reducing contingencies and wishful thinking. Costs need to be treated as important design parameters in the same way as the physical and performance parameters, such as loads, speeds, and so on.

If inadequate costing procedures are used, it is impossible to optimize costs and a 'non-optional' concept may be implemented unintentionally. It is a fact that the new approaches being taken in manufacturing have not yet been matched in the financial areas. The danger of this is that opportunities can be missed by accounting systems that have difficulty in putting a true cost on factors such as reduced lead times and improved quality.

The standard costing system found in most companies is not suitable for determining an accurate cost for the product. It records costs historically and then determines appropriate corrective actions. The system is more comfortable when dealing with direct labour costs, rather than those associated with the purchase of expensive capital equipment requiring a large investment. There is now a need for cost planning rather than cost reporting, particularly in an environment that uses such advanced, high-capital manufacturing equipment. In this environment, the processes can be predefined and the 'fixed' costs generally predetermined with the variable costs, except for materials, being comparatively small. Cost planning that takes into account future costs together with future activities and resources is crucial in ensuring that truer costs are established at the design stage.

If there is one message that comes through from all of the authors of papers in this part, it is to reject designs that are not designed for manufacture. It is not practical to reduce costs effectively, once designs go into production. The benefits of the multi-disciplinary approach advocated will increase creativity by stimulating discussion on potential trade-offs. While it is true that 'designing by committee' is generally unproductive, and designers must be given the ultimate responsibility for original designs, they cannot be expected to be knowledgeable about all aspects of manufacture. Therefore, the best designs will only be achieved following a disciplined team effort, in which designs and manufacture are tackled simultaneously.

References

Eaton R. J. (1987). Product planning in a rapidly changing world. *International Journal of Technology Management*, **2**(2), 183–9

Martin J. M. (1988). The final piece of the puzzle. *Manufacturing Engineering*, September, 46–51

2.1

Manufacturing by design

Daniel E. Whitney

Robotics and Assembly Systems Division, Charles
Stark Draper Laboratory, Inc., Cambridge, MA, USA

In many large companies, design has become a bureaucratic tangle, a process confounded by fragmentation, overspecialization, power struggles and delays. An engineering manager responsible for designing a single part at an automobile company told me recently that the design process mandates 350 steps – not 350 engineering calculations or experiments but 350 work-ups requiring 350 signatures. No wonder, he said, it takes five years to design a car; that's one signature every 3½ days.

It's not as if companies don't know better. According to General Motors executives, 70% of the cost of manufacturing truck transmissions is determined in the design stage. A study at Rolls-Royce reveals that design determines 80% of the final production costs of 2000 components (Corbett, 1986). Obviously, establishing a product's design calls for crucial choices – about materials made or bought, about how parts will be assembled. When senior managers put most of their efforts into analysing current production rather than product design, they are monitoring what accounts for only about a third of total manufacturing costs – the window dressing, not the window.

Moreover, better product design has shattered old expectations for improving cost through design or redesign. If managers used to think a 5% improvement was good, they now face competition that is reducing drastically the number of components and subassemblies for products and achieving a 50% or more reduction in direct cost of manufacture. And even greater reductions are coming, owing to new materials and materials-processing techniques. Direct labour, even lower cost labour, accounts for so little of the total picture that companies still focusing on this factor are misleading themselves not only about improving products but also about how foreign competitors have gained so much advantage.

In short, design is a strategic activity, whether by intention or by default. It influences flexibility of sales strategies, speed of field repair and efficiency of manufacturing. It may well be responsible for the company's future viability. I want to focus not on the qualities of products but on development of the processes for making them.

Converting a concept into a complex, high-technology product is an involved procedure consisting of many steps of refinement. The initial idea never quite works as intended or performs as well as desired. So designers make many modifications, including increasingly subtle choices of materials, fasteners, coatings, adhesives and electronic adjustments. Expensive analyses and experiments may be necessary to verify design choices.

In many cases, designers find that the options become more and more difficult; negotiations over technical issues, budgets and schedules become intense. As the design evolves, the choices become interdependent, taking on the character of an interwoven, historical chain in which later decisions are conditioned by those made previously.

Imagine, then, that a production or manufacturing engineer enters such detailed negotiations late in the game and asks for changes. If the product designers accede to the requests, a large part of the design may simply unravel. Many difficult and pivotal choices will have been made for nothing. Where close calls went one way, they may now go another; new materials analyses and production experiments may be necessary.

Examples of failure abound. One research scientist I know, at a large chemical company, spent a year perfecting a new process – involving, among other things, gases – at laboratory scale. In the lab the process operated at atmospheric pressure. But when a production engineer was finally called in to scale up the process, he immediately asked for higher pressures. Atmospheric pressure is never used in production when gases are in play because maintaining it requires huge pipes, pumps and tanks. Higher pressure reduces the volume of gases and permits the use of smaller equipment. Unfortunately, the researcher's process failed at elevated pressures, and he had to start over.

Or consider the manufacturer whose household appliance depended on close tolerances for proper operation. Edicts from the styling department prevented designs from achieving required tolerances; the designers wanted a particular shape and appearance and would not budge when they were apprised of the problems they caused to manufacturing. Nor was the machine designed in modules that could be tested before final assembly. The entire product was built from single parts on one long line. So each finished product had to be adjusted into operation – or taken apart after assembly to find out why it didn't work. No one who understood the problem had enough authority to solve it, and no one with enough authority understood the problem until it was too late. This company is no longer in business.

Finally, there was the weapon that depended for its function on an infrared detector, the first of many parts – lenses, mirrors, motors, power supplies, etc. – that were glued and soldered together into a compact unit. To save money, the purchasing department switched to a cheaper detector, which caused an increase in final test failures. Since the construction was glue and solder, bad units had to be scrapped. Someone then suggested a redesign of the unit with reversible fasteners to permit disassembly. But this time more reasonable voices prevailed. Reversible fasteners would have actually increased the weapon's cost and served no purpose other than to facilitate factory rework. Disassembly would not have been advisable because the unit was too complex for field repair. It was a single-use weapon – with a shelf life of five years and a useful life of 10 seconds. It simply had to work the first time.

Manufacturers can avoid problems like this. Let's look at a success story. One company I know wanted to be able to respond in 24 hours to worldwide orders for its electronic products line – a large variety of features in small-order batches. Engineers decided to redesign the products in modules, with different features in each module. All the modules are plug compatible, electrically and mechanically. All versions of each module are identical on the outside where assembly machines handle them. The company can now make up an order for any set of features by selecting the correct modules and assembling them, all of this without any human intervention, from electronic order receipt to the boxing of final assemblies.

In another company, a high-pressure machine for supplying cutting oil to machine tools requires once-a-day cleaning. Designers recently reconfigured the machine so that normal cleanout and ordinary repairs can be accomplished without any tools, thus solving some bothersome union work-rule problems.

There are no guarantees, of course, but the experiences of these companies illustrate how design decisions should be integrated, informed and balanced, and how important it is to involve manufacturing engineers, repair engineers, purchasing agents and other knowledgeable people early in the process. The product designer asks, 'What good is it if it doesn't work?' The salesperson asks, 'What good is it if it doesn't sell?' The finance person asks, 'What good is it if it isn't profitable?' The manufacturing engineer asks, 'What good is it if I can't make it? The team's success is measured by how well these questions are answered.

■ The design team and its task

Multi-functional teams are currently the most effective way known to cut through barriers to good design. Teams can be surprisingly small – as small as four members, though 20 members is typical in large projects – and they usually include every specialty in the company. Top executives should make their support and interest clear. Various names have been given to this team approach, like 'simultaneous engineering' and 'concurrent design'. Different companies emphasize different strengths within the team. In many Japanese companies, teams like this have been functioning for so long that most of the employees cannot remember another way to design a product.

Establishing the team is only the beginning, of course. Teams need a step-by-step procedure that disciplines the discussion and takes members through the decisions that crop up in virtually every design. In traditional design procedures, assembly is one of the last things considered. My experience suggests that assembly should be considered much earlier. Assembly is inherently integrative. Weaving it into the design process is a powerful way to raise the level of integration in all aspects of product design.

A design team's charter should be broad. Its chief functions include:

(1) Determining the character of the product, to see what it is and thus what design and production methods are appropriate.

(2) Subjecting the product to a product function analysis, so that all design decisions can be made with full knowledge of how the item is supposed to work and all team members understand it well enough to contribute optimally.

(3) Carrying out a design-for-producibility-and-usability study to deter-mine if these factors can be improved without impairing functioning.

(4) Designing an assembly process appropriate to the product's particular character. This involves creating a suitable assembly sequence, identifying subassemblies, integrating quality control and designing each part so that its quality is compatible with the assembly method.

(5) Designing a factory system that fully involves workers in the production strategy, operates on minimal inventory, and is integrated with vendors' methods and capabilities.

☐   **The product's character**

Clearly it is beyond the scope of this article to establish by what criteria one judges, develops or revamps the features of products. Recently in these pages, David A. Garvin (1987) has analysed eight fundamental dimensions of product quality; and John R. Hauser and Don Clausing (1988) have explored ways to communicate to design engineers the dimensions consumers want – in the engineers' own language.

Character defines the criteria by which designers judge, develop or revamp product features. I would only reiterate that manufacturing en-gineers and others should have something to say about how to ensure that the product is field repairable, how skilled users must be to employ it successfully, and whether marketability will be based on model variety or availability of future add-ons.

An essential by-product of involving manufacturing, marketing, purchasing and other constituencies in product conception, moreover, is that diverse team members become familiar enough with the product early in order to be able to incorporate the designers' goals and constraints in their own approaches. As designers talk with manufacturing or field-service reps, for example, they can make knowledgeable corrections. ('Why not make that part out of plastic? I know a low-cost source.' 'Because the temperature there is 1000°; plastic will vaporize.' 'Oh.')

☐ Product function analysis

This used to be the exclusive province of product designers. But now it is understood that to improve a product's robustness, to 'design quality in' in Genichi Taguchi's good phrase, means thoroughly understanding a product's function in relation to production methods. Product designers and manufacturing engineers used to try to understand these relations by experience and intuition. Now they have software packages for modelling and designing components to guide them through process choices – software that would have been thought fantastic a generation ago.

Recently I worked on a product containing delicate spinning parts that had to be dynamically balanced to high tolerances. In the original design, partial disassembly of the rotating elements after balancing was necessary before the assembly could be finished, so the final product was rarely well balanced and required a lengthy adjustment procedure. Since total redesign was not feasible, the team analysed the reassembly procedure solely as it pertained to balance and concluded that designers needed only to tighten various tolerances and reshape mating surfaces. Simple adjustments were then sufficient to restore balance in the finished product.

Another important goal of product function analysis is to reduce the number of parts in a product. The benefits extend to purchasing (fewer vendors and transactions), manufacturing (fewer operations, material handlings and handlers) and field service (fewer repair parts).

When a company first brings discipline to its design process, reductions in parts count are usually easy to make because the old designs are so inefficient. After catching up, though, hard, creative work is necessary to cut the parts count further. One company I know saved several million dollars a year by eliminating just one subassembly part. The product had three operating states: low, medium and high. Analysis showed that the actions of one part in the original design always followed or imitated the actions of two others. Designers eliminated the redundant part by slightly altering the shapes of the other two parts.

This change could never have been conceived, much less executed, if the designers hadn't had deep knowledge of the product and hadn't paid attention to the actions underlying its engineering.

☐ Design for producibility

Recently, a company bragged to a business newsweekly about saving a mere $250 000 by designing its bottles for a new line of cosmetics to fit existing machines for filling, labelling and capping. This plan seems so obvious, and the savings were so small as compared with what is possible, that the celebration seemed misplaced. But it's a better outcome than I remember from my first job with a drug company. It spent a fortune to have a famous industrial designer create new bottles and caps for its line. They were triangular in cross-section and teardrop shaped, and they would not fit either existing machines or any new ones we tried to design. The company eventually abandoned the bottles, along with the associated marketing campaign.

Obviously, nothing is more important to manufacturing strategy than designing for the production process. In the past, this has meant designing for manufacturing and assembly, and value engineering, which both strive to reduce costs. But now we have to go beyond these goals.

To take the last point first, value engineering aims chiefly to reduce manufacturing costs through astute choice of materials and methods for making parts. Does the design call for metal when a ceramic part will do? If metal, should we punch it or drill it? Value engineering usually comes into play after the design is finished, but the thoroughness we seek in design can be achieved only when decisions are made early.

Moreover, design for producibility differs from design for assembly, which typically considers parts one by one, simplifies them, combines some to reduce the parts count or adds features like bevels around the rims of holes to make assembly easier. Valuable as it is, the process cannot achieve the most fundamental improvements because it considers the product as a collection of parts instead of something to satisfy larger goals, such as reducing costs over the product's entire life cycle.

Nippondenso's approach vividly illustrates how an overriding strategy can determine a product's parts and the production process. The Delco of Japan, Nippondenso builds such car products as generators, alternators, voltage regulators, radiators and anti-skid brake systems. Toyota is its chief customer. Nippondenso has learned to live with daily orders for thousands of items in arbitrary model mixes and quantities.

The company's response to this challenge has several components:

- The combinatorial method of meeting model-mix production requirements.
- In-house development of manufacturing technology.

- Wherever possible, manufacturing methods that don't need jigs and fixtures.

The combinatorial method, carried out by marketing and engineering team members, divides a product into generic parts or subassemblies and identifies the necessary variations of each. The product is then designed to permit any combination of variations of these basic parts to go together physically and functionally. (If there are six basic parts and three varieties of each, for example, the company can build $3^6 = 729$ different models.) The in-house manufacturing team co-operates in designing the parts, so the manufacturing system can easily handle and make each variety of each part and product.

Jigless production is an important goal at this point, for obvious reasons. Materials handling, fabrication and assembly processes usually employ jigs, fixtures, and tools to hold parts during processing and transport;

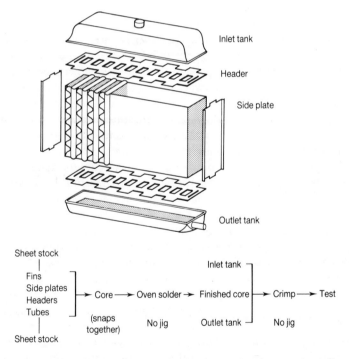

Figure 1 How a radiator is made – the combinatorial, jigless method. Except for final testing, this radiator is fabricated entirely without manual labour. Radiators differ by core length, width and depth. They are available in various sizes and offer many heat-transfer capacities. Without changing part sizes, a designer can program different shapes for fins and diameters for tubes, thus allowing the same production system to achieve new heat-transfer capacities.

the jigs and fixtures are usually designed specifically to fit each kind of part, to hold them securely. When production shifts to a different batch or model, old jigs and tools are removed and new ones installed. In mass-production environments, this changeover occurs about once a year.

In dynamic markets, however, or with just-in-time, batches are small, and shifts in production may occur hourly – even continually. It may be impossible to achieve a timely and economical batch-size-of-one production process if separate jigs are necessary for each model. Nippondenso's in-house manufacturing team responds to this problem by showing how to design the parts with common jigging features, so that one jig can hold all varieties, or by working with designers to make the product snap or otherwise hold itself together so that no clamping jigs are needed.

By cultivating an in-house team, Nippondenso also solves three difficult institutional problems. First, the company eliminates proprietary secrecy problems. Its own people are the only ones working on the design or with strategically crucial components. Second, equipment can be delivered without payment of a vendor's mark-up, thus reducing costs and making financial justification easier. Third, over the years the team has learned to accommodate itself intuitively to the company's design philosophy, and individual team members have learned how to contribute to it. Designers get to know each other too, creating many informal communication networks that greatly shorten the design process. Shorter design periods mean less lead time, a clear competitive advantage. (It is worth noting that many Japanese companies follow this practice of designing much of their automation in-house, while buying many product components from outside vendors. American companies usually take the opposite tack: they make many components and buy automation from vendors.)

Nippondenso uses combinatorial design and jigless manufacturing for making radiators (Figure 1). Tubes, fins, headers and side plates comprise the core of the radiator. These four snap together, which obviates the need for jigs, and the complete core is oven soldered. The plastic tanks are crimped on. The crimp die can be adjusted to take any tank size while the next radiator is being put in the crimper, so radiators can be processed in any model order and in any quantity. When asked how much the factory cost, the project's chief engineer replied, 'Strictly speaking, you have to include the cost of designing the product.' A factory isn't just a factory, he implied. It is a carefully crafted fusion of a strategically designed product and the methods for making it.

Without a guiding strategy, there is no way to tell what suggestions for improvement really support long-range goals. Some product design techniques depend too much on rules, including rule-based systems stemming from expert systems. These are no substitute for experienced people. Volkswagen, for example, recently violated conventional ease-of-assembly rules to capture advantages the company would not otherwise have had.

In the company's remarkable Hall 54 facility in Wolfsburg, Germany, where Golfs and Jettas go through final assembly, robots or special machines perform about 25% of the final-stage steps. (Before Hall 54 began functioning, Volkswagen never did better than 5% (Hartwich, 1985).)

To get this level of automation, VW production management asked to examine every part. It won from the board of directors a year-long delay in introducing the new models. Several significant departures from conventional automotive design practices resulted, the first involving front-end configuration. Usually, designers try to reduce the number of parts. But VW engineers determined that at a cost of one *extra* frame part the front of the car could be temporarily left open for installation of the engine by hydraulic arms in one straight, upward push. Installing the engine used to take a minute or longer and involved several workers. VW now does it unmanned in 26 seconds.

Another important decision concerned the lowly screw. Purchasing agents usually accept the rule that low-cost fasteners are a competitive edge. VW engineers convinced the purchasing department to pay an additional 18% for screws with cone-shaped tips that go more easily into holes, even if the sheet metal or plastic parts were misaligned. Machine and robot insertion of screws thus became practical. Just two years later, so many German companies had adopted cone-pointed screws that their price had dropped to that of ordinary flat-tip screws. For once, everyone from manufacturing to purchasing was happy.

☐   **Assembly processes**

Usually assembly sequence is looked at late in the design process when industrial engineers are trying to balance the assembly line. But the choice of assembly sequence and the identification of potential subassemblies can affect or be affected by – among other factors – product-testing options, market responsiveness and factory-floor layout. Indeed, assembly-related activities with strategic implications include: subassemblies, assembly sequence, assembly method for each step and integration of quality control.

Imagine a product with six parts. We can build it many ways, such as bottom up, top down or from three subassemblies of two parts each. What determines the best way? A balance of many considerations: construction needs, like access to fasteners or lubrication points; ease of assembly (some sequences may include difficult part matings that risk damage to parts); quality control matters, like the operator's ability to make crucial tests or easily replace a faulty part; process reasons, like ability to hold pieces accurately for machine assembly; and, finally, production strategy advantages, like making subassemblies to stock that will be common to many models, or that permit assembly from commonly available parts.

Again, software now exists to help the designer with the formidable problem of listing all the possible assembly sequences – and there can be a lot, as many as 500 for an item as simple as an automobile rear axle. It would be impossible for a team to attack so complex a series of choices without a computer design aid to help, according to a pre-established hierarchy of goals like that just discussed – access to lubrication points, etc. Another virtue of this software is that it forces the team to specify choices systematically and reproducibly, for team members' own edification but also in a way that helps justify design and manufacturing choices to top management.

Consider then, automatic transmissions, complex devices made up of gears, pistons, clutches, hydraulic valves and electronic controls. Large transmission parts can scrape metal off smaller parts during assembly, and shavings can get into the control valves, causing the transmission to fail the final test or, worse, fail in the customer's car. Either failure is unacceptable and terribly expensive. It is essential to design assembly methods and test sequences to pre-empt them.

With respect to assembly machines and tooling, manufacturers should consider the following questions:

- Can the product be made by adding parts from one direction, or must it be turned over one or more times? Turnovers are wasted motion and costly in fixtures.

- As parts are added in a stack, will the location for each subsequent part drift unpredictably? If so, automatic assembly machines will need expensive sensors to find the parts, or assembly will randomly fail, or parts will scrape on each other too hard.

- Is there space for tools and grippers? If not, automatic assembly or testing aren't options.

- If a manufacturing strategy based on subassemblies seems warranted, are the subassemblies designed so they do not fall apart during reorientation, handling or transport?

There are clear advantages to combining consideration of these assembly procedures and/or quality control strategy with design. Designers who anticipate the assembly method can avoid pitfalls that would otherwise require redesign or create problems on the factory floor. They can also design better subassemblies to meet functional specifications – specifications that will be invaluable when the time comes to decide whether to take bids from outside vendors or make the part on the company's own lines, specifications that will determine how to test the subassembly before adding it to the final product.

Designers concerned about assembly must ask:

- What is the best economic combination of machines and people to assemble a certain model-mix of parts for a product line (given each machine's or person's cost and time to do each operation, plus production-rate and economic-return targets)?

- How much time, money, production machinery or in-process inventory can be saved if extra effort is put into design of the product, its fabrication and assembly processes, so that there are fewer quality control failures and product repairs? A process that yields only 80% successful assemblies on the first try may need 20% extra capacity and inventory – not to mention high-cost repair personnel – to meet the original production goals.

- Where in the assembly process should testing take place? Considerations include how costly and definitive the test is, whether later stages would hide flaws detectable earlier, and how much repaired or discarded assemblies would cost.

These are generic problems; they are hard to answer, and they too are stimulating the development of new software packages. This new software enhances the ability of manufacturing people to press their points in (often heated) debates about design. Hitherto, product designers, more accustomed to using computer modelling, have had somewhat of an upper hand.

☐   **Factory system design**

Many features of good product design presuppose that machines will do the assembly. But automation is not necessary to reap the benefits of strategic design. Indeed, sometimes good design makes automatic assembly unnecessary or uneconomic by making manual assembly so easy and reliable. Regardless of the level of automation, some people will still be involved in production processes, and their role is important to the success of manufacturing.

Kosuke Ikebuchi, general manager of the General Motors–Toyota joint venture, New United Motors Manufacturing Inc, NUMMI, believes that success came to his plant only after careful analysis of the failures of the GM operation that had preceded it: low-quality parts from suppliers, an attitude that repair and rework were to be expected, high absenteeism resulting in poor workmanship, and damage to parts and vehicles caused by transport mechanisms (1986). The assembly line suffered from low efficiency because work methods were not standardized, people could not repair their own equipment and equipment was under-utilized. Excess inventory, caused by ineffective controls, was another problem. Work areas were crowded. Employees took too much time to respond to problems.

NUMMI's solutions focused on the Jidoka principle – quality comes first. According to NUMMI's factory system today, workers can stop the line if they spot a problem; the machinery itself can sense and warn of problems. Two well-known just-in-time methods of eliminating waste – the kanban system of production control and reductions in jig and fixture change times – are important to NUMMI's manufacturing operation.

But lots of other things also contribute to this plant's effectiveness: simplified job classifications, displays and signs showing just how to do each job and what to avoid, self-monitoring machines. NUMMI has obtained high-spirited involvement of the employees, first by choosing new hires for their willingness to co-operate, then by training them thoroughly and involving them in decisions about how to improve the operations.

■ Design means business

The five tasks of design bring us back to the original point. Strategic product design is a total approach to doing business. It can mean changes in the pace of design, the identity of the participants and the sequence of decisions. It forces managers, designers and engineers to cross old organizational boundaries, and it reverses some old power relationships. It creates difficulties because it teases out incipient conflict, but it is rewarding precisely because disagreements surface early, when they can be resolved constructively and with mutual understanding of the outcome's rationale.

Strategic design is a continual process, so it makes sense to keep design teams in place until well after product launching when the same team can then tackle a new project. Design – it must be obvious by now – is a company-wide activity. Top management involvement and commitment are essential. The effort has its costs, but the costs of not making the effort are greater.

References

Corbett J. (1986). Design for economic manufacture. *Annals of CIRP*, **35**(1), 93

Garvin D. A. (1987). Competing on the eight dimensions of quality. *Harvard Business Review*, November/December, 101

Hartwich E. H. (1985). Possibilities and trends for the application of automated handling and assembly systems in the automotive industry. In *International Congress for Metalworking and Automation*, Hannover, West Germany, 126

Hauser J. R. and Clausing D. (1988). The house of quality. *Harvard Business Review*, May/June, 63

Ikebuchi, K. (1986). *Future Role of Automated Manufacturing Conference*, New York University, unpublished remarks.

Acknowledgements

I am indebted to my colleagues James L. Nevins, Alexander C. Edsall, Thomas L. De Fazio, Richard E. Gustavson, Richard W. Metzinger, Jonathan M. Rourke and Donald S. Seltzer for their contributions to this article. We have worked together for many years developing the ideas expressed here.

2.2
How design can boost profit

John Corbett
Cranfield Institute of Technology

An increasing number of manufacturing companies are realizing that to compete effectively, particularly with the low-wage countries and Japan, they must reduce the cost of their products while improving, or at least maintaining their quality.

Major reductions in product costs can be achieved by an increase in automation and designing for economic manufacture, which implies designing for automated manufacturing methods. However, in many cases these areas have been neglected because the vast majority of designers have concentrated their efforts on achieving product function, paying little attention to production, and even less to the problems of assembly. This has happened in spite of the high cost of assembly which can typically account for between 40 and 60% of the total cost of production. Too often products have been designed which can only be produced and assembled at an unacceptable cost, which in turn limits their market appeal.

This happens because designers do not appreciate the importance of designing for economic manufacture, and their work is often subjected to undue pressures caused by the need to launch new products in unreasonably short time-scales. What is not sufficiently understood is that additional care in the design office is likely to reduce both development time and costs, and lead to the launch of a better, lower-cost product within similar time-scales.

The greatest impact on assembly costs can usually be made by reducing the number of parts, particularly by eliminating those which are unnecessary, or by integrating one part with another as illustrated in Figure 1. The integration of parts may be achieved by using new materials or new manufacturing techniques (like outsert moulding). Although this generally results in increased complexity of an individual component, reducing the

This paper is reprinted by permission of the author from *Eureka Transfers Technology*, May 1987, pp. 59–65.

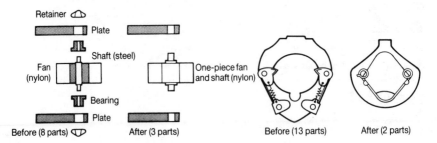

Figure 1   Reduce assembly costs by reducing parts.

number of parts and the number of production stages is an important factor which the designer should consider in the early stages of product design. The basic alternatives open to the designer in reducing the cost of assembly are either to avoid certain assembly operations altogether or to simplify them. To achieve these aims an investigation should be undertaken in which the designer seeks answers to the following simple questions:

- Does one part move relative to another?
- Can adjacent parts be of the same material?
- How much will integration affect production and assembly costs?

The results of this investigation will frequently identify those parts which are 'suspect' as individual entities, and if the activity is carried out with sufficient rigour it is possible to achieve extremely large cost savings.

The number of parts may also be reduced by limiting the required number of adjustments, some of which may, on closer inspection, be difficult or even impossible to do in assembly. In any event, reducing the number of adjustments can minimize the use of threaded fasteners, tools and fixtures. Designers should try to ensure that the most critical dimensions are contained within a single element or part, in order to minimize the build-up of tolerances and the subsequent need for adjustment. A fundamental approach is to:

- list all known functions;
- rank dimensions in order of importance;
- list all handling requirements, and indicate how replacement units will be fitted in the field.

It must be realized, however, that there is no 'cook book' or single solution for a designer's problem, and trade-offs are invariably required. For example, the integration of parts may, in some instances, escalate costs to such an extent as to nullify assembly advantages, and quality may also suffer if insufficient care is taken during the analysis stage. Materials, geometry, dimensions, surface finish, etc. are all important factors regarding quality because the assembly process itself will require the utilization of certain surfaces. Positive location surfaces and features, such as centring spigots, dowel pins and tenons, can remove tedious alignment and measuring operations.

Checklist 1   Design for assembly

Minimize the number of:
- parts anf fixings;
- design variants;
- assembly movements;
- assembly directions.

Provide:
- suitable lead–in chamfers;
- automatic alignment;
- easy access for locating surfaces;
- symmetrical parts, or exaggerate asymmetry;
- for simple handling and transportation.

Avoid:
- visual obstructions;
- simultaneous fitting operations;
- parts which will tangle or 'nest';
- adjustments which affect prior adjustments;
- the possibility of assembly errors.

There are many alternative methods of fastening. These can fall into two general categories: permanent (weldments, adhesives, etc.) or temporary (screws, circlips, etc.). In addition, a method which can fit either category is the use of 'snap-in' devices which achieve assembly and retain attachment by utilizing the elastic properties of the component or device itself.

The use of adhesives is increasing, although it is still not possible to achieve universal agreement regarding the relative merits of adhesive bonding and mechanical fastening. The advantages of adhesives are:

- generally less costly;
- minimum of component distortion;
- appearance not affected;
- reduced weight;
- greater damping of mechanical vibrations;

- wide area of stress distribution;
- galvanic corrosion minimized.

The advantages of mechanical fastening are:

- generally stronger;
- some can be reused;
- useful over a wide range of temperatures;
- dual function, for instance component mounting;
- well-established technology with no special training requirement.

With the ever growing number of adhesives being developed it is important that the designer avoids selecting one by himself, but, rather, seeks advice from specialist vendors. The methods of adhesive application and cure should also be considered as integral parts of the selection process. The main aspects to consider in the selection of the right adhesive are the performance requirements, the substrate and the joint design. A knowledge of the performance requirement can identify suitable adhesives, such as the potential of hot melts. Evaluation tests are recommended prior to final selection because suppliers can often be over-enthusiastic regarding their own product's suitability for a particular application. In order to achieve a strong and consistent adhesive bond for reliable assembly it is necessary to prepare and 'wet' the surface of the substrate uniformly.

Joints must be designed for adhesive bonding so that a uniform stress distribution is obtained, and the basic stresses are in shear or tension, with any peel stress being minimized (Figure 2).

It is important that designers are aware of the methods of assembly to be adopted for their products. Manual operators have the advantage of 'feel', intuition and judgement, and can thus accomplish assembly operations which are very difficult, or in some cases impossible, for a robot or automated assembly workstation. The assembly area has, in recent years, proved to be one of the fastest growing fields for the application of robots. However, a major limitation associated with robotic assembly is its slow production rate when compared to a dedicated assembly machine.

Considerable time and financial resources are needed to solve the large number of difficulties associated with the automation of the assembly area. Nevertheless some industries have no alternative if they wish to remain competitive. Product designs must therefore be thoroughly reviewed for automation so that assembly movements are kept simple and that sophisticated, expensive workstations with high failure rates are avoided.

A substantial amount of automation is being undertaken in the electronics and domestic appliance industries, both of which employ high-volume production but require frequent changes in product design, thus making a 'fixed' assembly line undesirable.

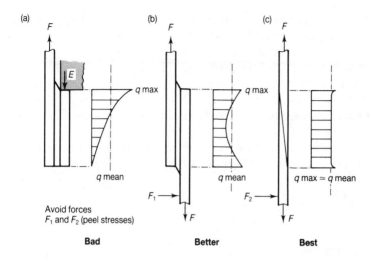

Figure 2 Shear stress (q) distribution in adhesive joints ignoring bending and peel stresses.

There is little evidence to date of universal, or truly flexible, assembly machines. It is therefore desirable to combine the design of the product with the design of the specialized equipment required for its assembly. Designers have a demanding task and it is important that they have a clear under-standing of the general principles of designing for assembly. The most successful automation applications have evolved when both the product and its automated assembly equipment have been designed simultaneously. It is therefore of strategic importance for companies to consider the methods of assembly in the early stages of the design of new products.

Two major research programmes, which are now being adopted commercially, have spearheaded progress in designing for assembly, and in particular for automatic assembly. The programmes are the Assemblability Evaluation Method, AEM, developed by Hitachi in Japan, and the Design for Assembly, DFA, method, which was developed by Professor G. Boothroyd in the USA, with strong collaboration from the University of Salford in the UK.

Both methods give designers and production engineers an idea of the ease, or difficulty, associated with components, pointing out weaknesses regarding their ease of assembly. They work by assisting the designer in evaluating and quantifying the difficulty of assembly of parts by means of a points system which feeds information back to the designer. Impressive results have been claimed through the use of both AEM and DFA. The evaluation and subsequent iterations resulting from their use have proved to be very effective in improving product designs in their early stages.

Hitachi have given details of one of their products, a video tele-recorder, VTR, produced at 75 000 per month in which huge savings were made by using the AEM (Heginbotham, 1984). These savings were achieved through a significant reduction in the number of parts, the easier assembly of the remaining parts and by designing the automatic assembly system in conjunction with the design of the product. The company claims that the assembly line developed for the new VTR mechanism resulted in an 83% reduction in labour, a one-third reduction in assembly shop area and an 80% improvement in quality.

The application of AEM is said to be mandatory within Hitachi, which is not surprising if the results above are being achieved.

Checklist 2

Ideal components for automated assembly should:

- be consistent in shape and size;
- be free from burrs or flashing;
- require a minimum amount of orientation;
- have features which ensure correct orientation;
- be smooth in surface finish;
- be free from oil or swarf;
- be able to withstand high-speed mechanical handling without damage.

Professor Boothroyd's DFA method is primarily intended for mechanical parts and is not so suitable for pcb assembly applications. DFA has received considerable interest in the USA and has been tested by many large companies including Xerox, IBM, General Motors and 3M. In Europe it has been assessed by Philips in the Netherlands, who believe that it can give the company significant assistance (Boorsma, 1985). The company has described it as an important aid for designers in achieving parts which are easier to manufacture. Figure 3 shows one example in which the number of parts has been reduced from 16 to 7 and assembly times slashed by over 70%.

Philips use the DFA information to identify key assembly problems. Questions have to be answered for all items and assembly operations and the results are analysed. After analysing 30 products for various divisions – such as audio, video, lighting and domestic appliances – Philips have achieved encouraging results. For instance, number of parts reduced by 25%, assembly times reduced by 28%, number of operations reduced by 30% and parts which can be assembled automatically have increased by 200%.

DFA and AEM both involve more than just design rules such as 'ensure that a part is symmetrical' and 'avoid parts which can tangle or nest'. The majority of designers do not pay attention to design rules as such. The two methods described are made more practical, and complement design

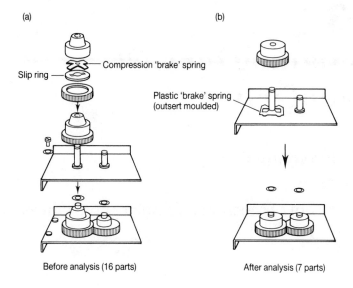

(a)

(b)

Compression 'brake' spring

Slip ring

Plastic 'brake' spring
(outsert moulded)

Before analysis (16 parts)

After analysis (7 parts)

Figure 3 Friction coupling before and after DFA.

rules by indicating the direction the designer should follow. They are used as a 'post-analysis' to highlight assembly problems and to encourage the generation of ideas to prevent such problems finding their way into production.

Clearly methods such as AEM and DFA should be investigated by companies in the UK, especially those who must compete with the above-mentioned companies in areas of price and quality, and who also need to increase their technological capability. However, before they can do this, companies must break down the barriers which cause the design of a product to be hindered by a lack of co-operation through unexpressed and misunderstood interests of groups within the company. In too many instances 'after-the-fact' engineering occurs with its inherent high costs caused through lost savings in production and a duplication of engineering effort which could be avoided if the job were to be done right first time.

References

Boorsma J. (1985). *Design for Assembly*. Eindhoven: Philips
Heginbotham W. B., ed. (1984). *Programmable Assembly*. IFS (Publications) Ltd

2.3
Designing for CNC manufacture

B. A. Hodgson and
G. Pitts

Department of Mechanical Engineering, University of Southampton, UK

■ Introduction

In order to achieve further significant savings in manufacturing industry, a very radical rethink of the manufacturing process is required. Included within 'manufacturing' is design as well as the forming or cutting of material. Significant effort has been expended in automating the latter process: capstans, copying lathes and milling machines, and multi–spindle automatic lathes are all significant achievements in this direction. Sales experience over the past decade indicates that customers require frequent changes to products – a rapid manufacturing response but with smaller batch sizes. This requirement has put traditional manufacturing industry in some difficulty since the machinery listed above takes a considerable effort in setting up and therefore long production runs are needed. An answer to this problem has emerged and makes considerable use of computer-controlled machines; the advantage of these is that they can be easily programmed, give very high repeatability and do not require long production runs to be economic. Continuing along the long–established group technology route, machine tools are grouped together to make small, specialist production cells, which are serviced with varying degrees of automation. This servicing takes care of component supply to and from the cell as well as comprehensive tool management. The transport of components and tools could be done

This paper is reprinted by permission of the Council of the Institution of Mechanical Engineers from *Proceedings of the Institution of Mechanical Engineers* (1989), Vol. 203, pp. 39–45 manually or, in the most sophisticated operations, by computer-controlled vehicles. Within a factory there would be several such cells, each based upon a particular limited family of component shapes. With a high degree of automation, such manufacturing cells would be termed *flexible manufacturing systems*, FMS; these would also take care of component transfers within the cell from one machine to another. Naturally FMSs are very capital intensive and require a sustained flow of work into them. Without this flow they are not viable, and so they tend to be concentrated with large manufacturers where the financial risk can be spread and where the variety and volume of work is more likely to be available.

The foregoing is very much a manufacturing solution to what is primarily a design-originated problem. It is well documented that the majority of costs are determined at the design stage so that no matter how good the jig and tool designer, the production engineer and all the other production support people, the fact remains that they can influence final production costs only minimally. This statement does assume that there is a degree of competence in all departments; it does not imply that poor-quality people can be allowed in the production department. If a radical approach is adopted at the front end, that is at the design stage, then the possible savings could be enormous. A recently completed project at Southampton University has concentrated on this area. The reason for this concentration was to seek alternatives to the current vogue of placing great emphasis on the manufacturing-led solutions, that is the FMS, and to make better use of stand-alone computer numerical-controlled, CNC, machine tools. The latter emphasis arose incidentally from other research projects, but the actual situation is that there are a very considerable number of these machines in use by all sectors of the manufacturing industry, and so a small improvement in capability would yield significant results taken nationally. Figure 1 shows the purchase of CNC machine tools by manufacturing industry in the UK. As a comparison the annual sales of manual machine tools is also shown; the volume of these sales is surprising when the trend has been assumed in the direction of CNC. However, the importance of the latter category must be determined by their significant production potential as compared to the manual machine tools. As a rough guide, production times have been reduced by between one-sixtieth and one-fortieth so that machining times that were once reckoned in hours can now be accomplished in as many seconds on CNC machines.

If designers can be encouraged to take a large step into the manufacturing area, then the major bugbear of manufacturing – that of transferring components between machine tools – can be minimized or even eliminated. This large step has long been the Holy Grail of all production engineers. This paper will discuss various techniques available to the designer so that single machining set-ups can be achieved while still preserving the hitherto primary objective of satisfying the functional requirements of the component. From this simple concept of changing the emphasis of design, many production and assembly benefits result.

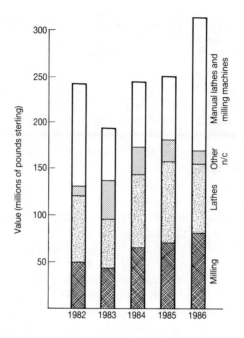

Figure 1   Use of CNC machine tools in the UK (from BSO Monitor PQ 3221, *Metal Working Machine Tools*).

■ Design strategy

This is based on two premises: that CNC machine tools can cope economically with certain complicated geometries and that component transfers between stand-alone machine tools dominate manufacturing times. The project addresses the metal removal sector of the manufacturing industry, and targets basic CNC machine tools. The capabilities of the more sophisticated four- and five-axes machines are not included in the study because they are in use by a small minority of manufacturers and they also benefit less from the objectives of this work, that is of optimizing low capital investment. This statement also includes turning centres with powered tooling installed in their turrets for the same reason.

In order that raw material can be placed on a CNC machine tool and then have sufficient geometry changes imposed on it such that the component is ready for final assembly, a considerable discipline is necessary within the design department. That discipline requires the designer to consider a list of priorities and then to proceed to the following item on the list with reluctance.

The designer should aim to achieve:

(a)   a single set-up on a machine tool, using one tool;

(b)   a single set-up on a machine tool, using the minimum number of tools;

(c)   the minimum number of set-ups on the same machine tool;

(d)   the minimum number of transfers between different machine tools.

Rules (c) and (d) recognize that not every component can be designed for single-set-up manufacture and therefore traditional design strategies have to be allowed.

■ Design examples

Several companies were approached at the start of the project and asked to contribute by submitting actual designs in the form of drawings. The intention was to examine as many designs as possible to cover a wide field of the engineering industry. In fact, companies as diverse as precision transducer manufacturers, scientific instrument builders and tensile test machine manufacturers collaborated. These designs were used to try and achieve single-set-up component machining. It was soon realized that tackling the project objective by taking the components, as designed, and refining the design was not going to be successful. As an alternative approach the design was started again from the concept stage using the original requirement specification.

The result of the revised design was that the 'new' components rarely looked anything like those they replaced. The techniques developed in the project permit details to be designed for single-set-up manufacture. Taken singly they may appear simple or even trivial. Taken together they could form a powerful contribution in reducing product costs.

In all of the techniques there is a common theme running which permits single-set-up manufacture. The golden rule that the designer needs to have constantly in mind is 'never deviate from the primary axis'. This rule has apparently different interpretations for turned and milled components; in fact the rule is the same for both types. Figures 2(a) and (b) and 3(a) and (b) show what is meant by the rule for the two processes. In both cases the *primary axis* refers to the axis of rotation which creates the relative cutting action between the workpiece and the tool.

The particular techniques described here are probably not original, but the requirement for designers to use them to achieve single-set-up machining is.

There are three techniques described which are shown in Figure 4. The first technique, Figure 4(a), is to remove a feature, in this case a cavity for a

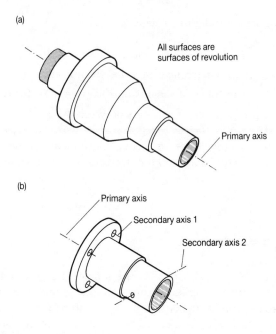

Figure 2 Turned component that (a) obeys and (b) does not obey the 'golden rule'.

filter, from a component requiring a second machining operation. The 'displaced' feature is then incorporated into the adjacent component.

Figure 4(b) illustrates a technique for avoiding cross-bores and at the same time permitting subassembly retention by deformable tags. This technique may not seem very sophisticated but it is certainly effective, low cost and has the advantage of reducing the number of components in the assembly since no separate fasteners are required. An additional benefit over conventional design, which would have required a cross-boring to accept the axle, is that the axle and wheel can be pre-assembled, and this subassembly can subsequently be dropped into position and retained. The alternative would have required the axle to be inserted part-way in the cross-bore, with the wheel placed in position so that the axle could be pressed through it and into the remaining part of the cross-bore. This would not be a particularly easy task and final assembly would almost certainly be longer.

The final technique to be described in this paper is shown in Figure 4(c). This is a small, turned component where the smaller bore can only be machined through the larger bore with a long, slender tool. The small bore is only 1 mm diameter and would not be practical from the direction described. Therefore a second operation would be required in which the piston body would be reversed in the lathe chuck so that the small hole could be drilled. However, if the component is split into two components, both can be

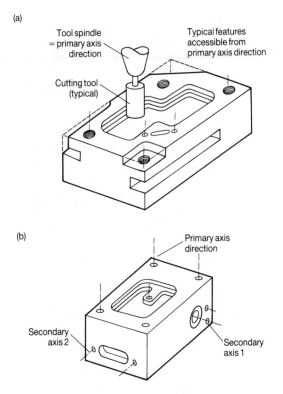

Figure 3   Milled component that (a) obeys and (b) does not obey the 'golden rule'.

machined in single operations and subsequently pressed together. There is another advantage in this approach since the small diameter part of the piston body is required to be titanium nitrided to achieve a very hard surface. This process is relatively expensive so that a reduction in component size will permit larger batches in the vacuum chamber. A subsidiary benefit of 'splitting' a component, as described above, is that if the small hole were required to be accurately positioned, with respect to a datum feature in the larger part of the piston body, then greater certainty would be obtained in achieving the tolerance than by reversing the component in the chuck.

■ Application to an assembly

Several good examples could have been chosen from a list covering tracker ball assemblies, fuze devices for naval ordinances, hybrid potentiometer and encoder for industrial robots, etc. As with the techniques outlined above,

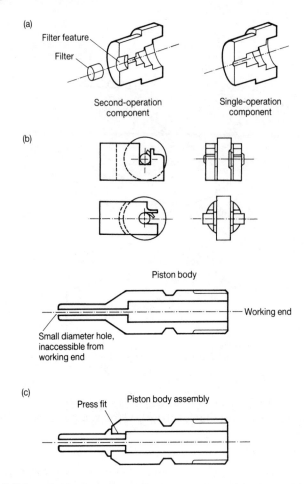

Figure 4 (a) Transfer features from a component to its neighbour; (b) Component retention by deformable tags; (c) Second-operation component made into two single-operation components.

space limits the scope of the presentation so that only one assembly example can be described. The example chosen is a low-speed, high-pressure, reciprocating pump used to circulate small quantities of solvent. The pump was originally conceived as a twin-piston, reciprocating machine with a stroke of 3 mm. The pistons are operated through cams and followers and are returned by spring action. The same operating principle is used in the 'new' design.

The original cylinder design is shown in Figure 5 where it can be seen that four machining set-up operations are required. The 'golden rule' has been broken, because the milling and drilling operations take place on different tool rotational axes, and the penalty is the multiple-machining

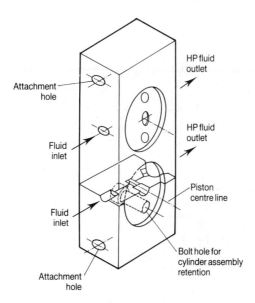

Figure 5 Original manifold for HP pump.

set-ups required. This may be compared to the new design in Figure 6 where the cylinders and pistons have been made from all-turned, single-operation components through which the working fluid passes axially, coinciding with each of the component's principal axes. Flexible pipes feed low-pressure fluid to the pistons. Both cylinder subassemblies screw into a rectangular, milled manifold whose components are also the result of single-operation machining. The manifold components are brazed, and to assist in its assembly operation, two guide pins have been introduced which may subsequently be removed if required. An alternative would be to have the pins as part of an assembly fixture.

The pistons are activated by followers whose construction is shown in Figure 7. An as-purchased pre-machined bar is used into which pockets are milled to accept the piston and the cam follower subassemblies.

■ Benefits to machine shop scheduling

The design method has a major benefit in the ease with which components can be scheduled through the manufacturing facility. If only one machine tool is required per component then the ability to predict when an assembly is going to be complete must have a higher confidence level than when multiple-machining set-ups are used. This is because different components, each in their own batches, have to queue at downstream machines for their

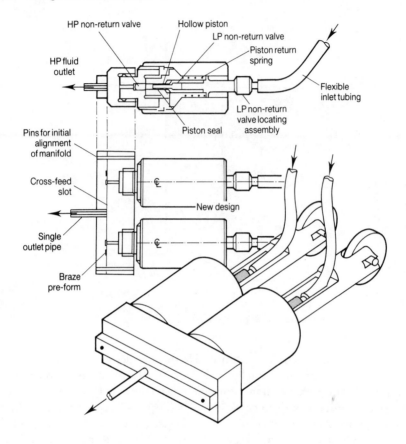

Figure 6   New design for manifold and cylinder.

next operation. The picture is further complicated by the fact that different types of machine have different metal removal rates, and each component will require more or less time on a particular machine. Material machining differences add another dimension to the variability.

Any small event that causes delay in upstream operations can have a disproportionately large effect in delaying downstream operations, which is why the prediction of assembly dates (and thus delivery of goods) has been open to considerable variability – usually on the late side rather than the early. The more machining operations required on a component, the worse the scheduling uncertainty becomes. An approach used in the past is to employ progress chasers to expedite particular component batches. This artificial intervention is very often self-defeating because priorities are arbitrarily changed according to 'who shouts loudest'. A common scenario in this situation is that nothing is delivered on time, not even those that start life with long lead times. Today, modern, computer-based, planning systems have

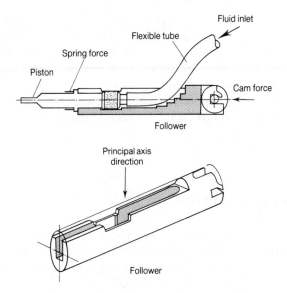

Figure 7   'New' cam follower design with retention tags.

potential for alleviating the scheduling problem to some degree. However, the premise that multiple-machining operations cause complications on the shop floor still holds true.

The design of the solvent pump, described above, has been costed by the pump manufacturer and shows a 10% saving on the original design. This costing was based on machine hourly rates which include a number of overhead charges. Some of these charges would be attributable to the scheduling and transfer of components between machine tools. If the 'new' design concept were to be taken on as a whole, then the chargeable hourly machine rates could be reduced so that the estimated saving of 10%, which is already significant, might increase to 12 or 15%.

■ Benefits to assembly

The majority of features machined into components are there for purposes of assembly as well as to create geometrical compliance with adjoining components. The drive to minimize multiple-machining set-ups results in components being machined from one side. It is this factor that provides the benefit, and means that components produced with the new design philosophy tend to have one side from which assembly operations always take place and a reverse which can be used for holding the item. The consequence of this is that simplified assembly fixtures and simple, automated assembly processes may be used. As an example it can be seen from Figures 6 and 7 that the subassemblies are built up from a single direction along their principal axes.

It will be realized that single-sided machining also benefits both numerical-controlled, NC, measuring machines and conventional surface plate inspection because the features have been formed from one direction and can therefore be measured from this single direction. It is expected that inspection probes and control software for NC measuring machines can be considerably simplified.

■ Providing the design link

In order that the average designer in industry can be guided to a definitive solution quickly, more than a limited collection of techniques is required. The solution arrived at must satisfy the functional requirements as well as being suitable for single-set-up machining. A system, in embryo form at least, has been devised which is based on a hierarchical or family tree structure for *function* (of a component). By considering six primary functions of a component at the highest level in the hierarchy – mechanical, structural, thermal, fluidic, tribological and aesthetic, – these may then be further subdivided until actual engineering solutions are suggested. Particular developments of part of the family tree are shown in Figures 8 and 9. The design methodology is aimed at stimulating the designer's imagination and broadening the approach so that several ways of fulfilling the required function can be considered. These can then be ranked according to their suitability for single-set-up machining: all of them will satisfy the functional requirement. The difficulty in devising such a system comes in formalizing the imaginative leap between the conventional way of designing components, which involves multiple set-ups, to the 'new' method. It is hoped that this problem will be addressed in a future project.

■ Conclusions

A method of design that considers the process planning aspects of manufacture by insisting that single-set-up manufacture is considered equal in importance to the function of a component must have the potential of achieving very large savings in manufacturing industry.

It has been demonstrated that the dual requirements of function and single-sided machining are possible in many instances. The fact remains, however, that it is easier for a designer not to be bothered with the additional task of achieving more economic manufacture. Therefore the proposed

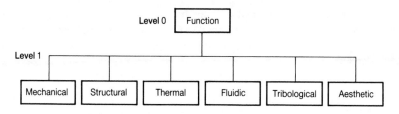

Figure 8 Proposed outline of the function diagram.

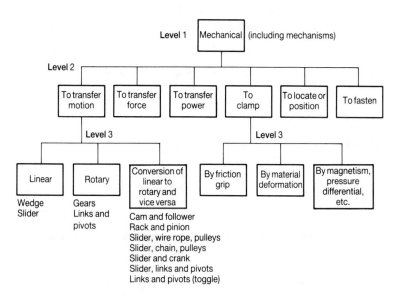

Figure 9 'Mechanical' subsection expanded to level 3.

project which will provide the design link between the initial requirement and the realization of actual components should proceed as soon as possible.

The benefits of being able to machine components on a single machine are manifold. The principal ones are firstly the ease of scheduling and the greater certainty of being able to predict delivery dates, the latter factor being particularly important with regard to the requirement to produce smaller batch sizes and to have more frequent product changes. Secondly, the components, being machined from a single direction, would be easier to inspect, resulting in less complex inspection set-ups which would therefore be quicker. Thirdly, assembly would be much simplified because the components would be stacked from the direction from which they had been machined.

While largely unquantified, the total cost saving to manufacturing industry is estimated to be very large. Cost estimates done to date, on a design that was by no means refined, indicate a saving of 10%; this is thought to be conservative.

Acknowledgements

The authors acknowledge the support of the ACME Directorate within the Science and Engineering Research Council under grant GR/C/90645 and the companies who supported the work.

2.4
Advanced castings in aircraft structures: a way to reduce costs

D. Mietrach
Messerschmitt-Bölkow-Blohm GmbH, Dept. TFB51,
Hünefeldstrasse 1-5, D-2800 Bremen 1, GFR

■ Introduction

Marked increases in the cost of raw materials, energy and labour in recent years have necessarily led to higher construction costs in the manufacture of airframes. The development and application of economical fabrication methods offers some promise of easing the situation. In the USA, for example, this tendency is becoming increasingly marked and considerable public resources are devoted to the promotion of extensive AMST structural programmes, advanced metallic structure technology (Faber, 1980). Figure 1 shows as an example the 'transverse frame' of the Boeing YC-14 as the reference component for the CAST program in the USA. In the conventional version, this 1400 × 2300 mm large component consists of approximately 400 detail parts and 2000 fasteners whereas the advanced design consists of only one sand casting of the alloy A357. It was demonstrated that the casting permits cost savings of 35% over the conventional component while having the same weight.

To remain competitive with the USA, the Europeans therefore concentrated on the increased use of new economic designs. As a result of this, the program for economic structure technologies using metals, WST-M (Mietrach, 1981 and 1982), funded by the Federal Ministry of Defence, aiming at a drastic reduction in aircraft construction costs by development and application of new economic production technologies, got under way at

This paper is reprinted by permission of the author from *MRS – Europe*, November 1985, pp. 201–11.

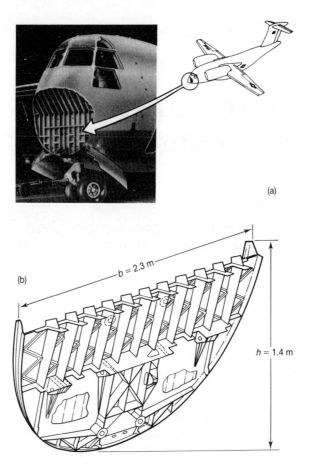

(a)

(b)

$b = 2.3$ m

$h = 1.4$ m

Figure 1   Transverse frame of the Boeing YC-14 as a reference component for the CAST program in the USA: (a) location of the component; (b) shape and main dimensions of aluminium casting version A357 (Mietrach, 1985).

MBB-UT/VFW in Bremen in 1978. The aluminium casting technology is an essential element within this WST-M program.

■ Advanced casting technology

☐ General

The casting technology is an extremely cost-effective production process since it provides the possibility of combining a multitude of material-intensive machined and sheet-metal parts (joined at great expense by riveting) in one casting. Moreover, it gives the designer greater freedom of design, for instance, casting of complicated undercuts.

An important point to bear in mind is that the design must be casting oriented, that is, the designer and caster should take up contact at an early stage of component design. Only thus will it be possible to make maximum use of all of the advantages of casting.

It is vital for the designer to be familiar with the characteristics and processes of different production procedures in order to use the inherent process advantages as effectively as possible (Mietrach, 1983).

☐ Casting processes

The lack of confidence in casting on the part of designers and stress engineers has been the main factor preventing the more extensive use of castings for parts of the primary structure.

But notable progress has been made on the casting technology sector making it possible also to use castings for primary structures that are subjected to high loads. Some points worth mentioning are low-pressure sand casting, casting in a vacuum, controlled solidification as well as the possibility of producing minimum wall thicknesses of 1.0 mm locally.

According to the present state of the art, three casting processes are particularly significant for use in the aerospace industry:

- Sand casting/low-pressure sand casting.
- Investment casting.
- Précial casting.

Of the casting procedures available (Table 1), sand casting is the most cost-effective solution for the production of large, complex components (Table 1). However, disadvantages over other procedures are that large tolerances (± 0.5 mm) and minimum wall thicknesses of not less than 2.0 mm are only possible at present. It is hoped to overcome these problems by local machining or selective chemical milling or the use of the low-pressure sand-casting process.

Investment castings are slightly more expensive than sand castings, but offer the possibility of producing thin wall thicknesses of approximately 1.5 mm at high tolerances (±0.15 mm). At present, components 500 × 800 × 1000 mm can be manufactured, this limit being set not by the process but by the available production facilities.

Précial casting is more expensive than sand and investment casting but makes it possible to produce components the size of sand castings with minimum wall thicknesses of 1.5 mm and high tolerances (± 0.2 mm). But because of the costs for this process and the progress of the other processes the précial foundry was closed.

Die castings are suitable for equipment or secondary structure only.

Table 1   Aluminium casting processes: overview.

	Investment casting	Premium sand casting	Conventional sand casting
Advantages	• Small tolerances • Small wall thickness	• High mechanical properties for all wall thickness • Small wall thickness • Small tolerances	• Large dimensions • Low costs • Local high mechanical properties by local cooling
Disadvantages	• Lower mechanical properties	• Higher costs for castings and tooling	• Larger tolerances • Larger wall thickness
Part size (maximum overall dimensions)	1000 × 800 × 500 mm	3600 × 2000 × 1000 mm	5000 × 1500 × 1500 mm
Wall thickness, depends on: • outer dimensions • part geometry • alloy	$T_{min} =$ 1.5 ± 0.15 mm	$T_{min} =$ 1.8 ± 0.4 mm 0.2 (Local: 1.5 ± 0.2 mm)	$T_{min} =$ 2.5 ± 0.5 mm (Local: 2 ± 0.5 mm)
Tolerances (D = considered dimension)	According to countries standard For example: VDG-P690 for Germany	$T_{min} = \pm$ (0.4 + 1.500/ 1000 mm	According to countries standard For example: DIN 1688 for Germany
Surface roughness	$R_a = 3.2\,\mu m$	$R_a = 3.2\text{–}6.4\,\mu m$	$R_a = 6.4\text{–}12\,\mu m$

☐   **Aluminium casting materials**

Aluminium alloys used for casting (Table 2) are characterized particularly by good mould-filling capabilities. The standard casting alloy A357 containing silicone (G-AlSi7Mg0,6), a further development of A356, has particularly good casting features with strength values of $R_m = 330\,N/mm^2$ (Premium/ sand casting). In cases where very high strength requirements have to be met, the silver alloy, highly cupriferous material A201 (G-AlCu3Ag) is particularly appropriate for sand and investment casting. This material has strength

Table 2 Casting alloys: Mechanical properties.

		Material specification		Mechanical properties (specimen, cut from castings)						Wall thickness (mm)
				Designated areas			Undesignated areas			
		Spec. no.	Short name	UTS N/mm²	YS N/mm²	Elong. %	UTS N/mm²	YS N/mm²	Elong. %	
Al alloys	Investment casting	3.2374 T6 (A356)	Al Si7 Mg 0.3	265	195	4	265	195	4	≤3
		3.2384 T6 (A357)	Al Si7 Mg 0.6	310	250	5	290	230	3	≤3
		A201 (AMS 4229)	Al cu 4.5 Ag 0.7 Mn 0.3 Mg 0.25 Ti 0.25	414*	345*	3*	386*	331*	1.5*	
	Premium/ sand casting	3.2374 T6 (A356)	Al Si7 Mg 0.3	270	200	4	230	190	2	≤18
		3.2384 T6 (A357)	Al Si7 Mg 0.6	330	270	5	305	240	3	≤20
		A201 (AMS 4229)	Al Cu 4.5 Ag 0.7 Mn 0.3 Mg 0.25 Ti 0.25	414*	345*	3*	386*	331*	1.5*	
Ti alloys	Invest. casting Rammed graphite	3.7264.1	Ti 6 Al 4 V	880	815	5	880	815	5	≤25

*According to AMS 4229.

73

values of approx. $R_m = 414\,\mathrm{N/mm^2}$ but is sensitive as regards thermal cracks during casting.

■ Costs

☐ General

A primary driving factor in the utilization of castings is cost reduction. Many studies have shown that castings can be extremely cost effective for particular applications. The two primary application areas are:

(1) Replacement of components which involve assembly of numerous details.

(2) Replacement of components requiring extensive machining.

However, there is no firm fixed rule on when or where a casting should be used. The economics will be affected by the casting process selected, production volume, material and quality requirements.

But most importantly the economics are affected by how good a job the designer does in designing the component to be cast. The cost of a casting can easily be doubled by factors controlled by the designer. It is highly recommended that the designer work with one or more foundries in developing his design. The foundry should make suggestions during the conceptional design and again during detail design (not after design is complete). The importance of this in getting cost-effective, high-quality castings cannot be over-emphasized.

Castings are not a panacea for all aircraft components. They should be used where it makes sense to use them. The following examples show two aircraft applications.

☐ Centre NIB structure

The NIB, located in the fixed wing area of the MRCA Tornado, is a primary component which is subjected to very high dynamic loads. The series version of this component consists of 15 machined and sheet-metal parts and 164 fasteners (Figure 2). The investment/précial casting, made of alloy A357, consists of only one part.

Extensive investigations and value analyses have shown that considerable cost savings are possible with the casting. Thanks to further design modifications and to the close contact with the foundries, the original target data of 15% weight savings over the series component at the same cost have

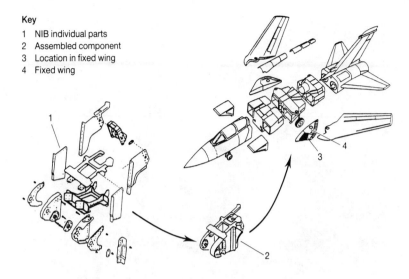

Key
1   NIB individual parts
2   Assembled component
3   Location in fixed wing
4   Fixed wing

Figure 2   The NIB, part of the MRCA Tornado primary structure, prior to the conversion.

Table 3   Weight and cost calculation for the NIB centre structure as investment/précial casting.

	Old: *Many single parts*	*New:* *Casting-* *version*		*Weight* *(%)*	*Costs* *(%)*
Number of parts	6 Machined parts 9 Sheet metal parts 164 Fasteners	1 casting	Goal	−15	±0
Weight (kg)	2.0	1.6	Achieved	−20	−25

been exceeded very notably with 20% weight savings and 25% cost savings (Table 3).

The design of the cast version was such that wall thicknesses of only 1.6 mm with high tolerances were achieved in large areas. Figure 3 shows the casting. The inspections and checks performed at the manufacturer's (foundry) and buyer's (MBB–UT, Bremen) in compliance with the test instructions served to verify the quality of the casting and also to draw conclusions as to the reproducibility of the delivered components.

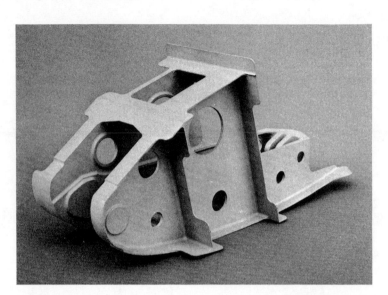

Figure 3 NIB: centre structure (military aircraft) investment casting (material: A357); size: 500 × 190 × 300 mm).

To sum up, it can be said that the test results meet the requirements according to the test instructions. More data concerning this example are published in Mietrach and Weilke (1982).

☐ **Intake floor**

The MRCA Tornado 'intake floor' is a representative example of a component with a complex geometry for a large investment casting or medium-sized sand casting, produced with the aluminium casting alloy A357. The component which belongs to the primary structure and is subjected to high loads is located in the forward engine air intake and bears the internal pressure loads of the intake.

At present, this component consists of an expensive mixed design comprising 22 machined/sheet-metal parts and approx. 400 fasteners (Figure 4). The cast version consists of one part only.

First value analyses have shown that, by comparison with the conventional design and at the same weight, cost savings of approx. 45% are possible. In fact, cost reductions of more than 60% can be achieved by modifying the design, making it more castable and using the low-pressure, sand–casting procedure (Figure 5). Figure 6 shows the cast intake floor.

Qualification tests performed at the manufacturer's and buyer's proved compliance with the mechanical and technological properties defined

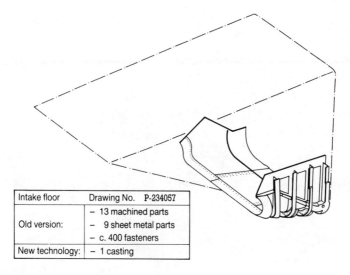

Intake floor	Drawing No. P-234057
Old version:	− 13 machined parts − 9 sheet metal parts − c. 400 fasteners
New technology:	− 1 casting

Figure 4   Exemplary component: MRCA Tornado intake floor.

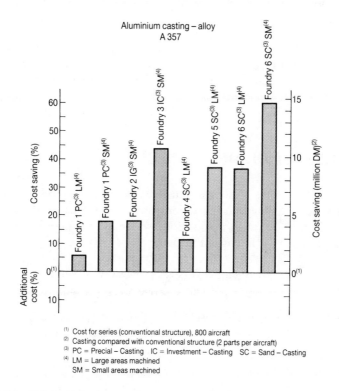

(1) Cost for series (conventional structure), 800 aircraft
(2) Casting compared with conventional structure (2 parts per aircraft)
(3) PC = Precial – Casting   IC = Investment – Casting   SC = Sand – Casting
(4) LM = Large areas machined
    SM = Small areas machined

Figure 5   MRCA intake floor. Cost comparison with basic component under series-production conditions for various casting processes/foundries.

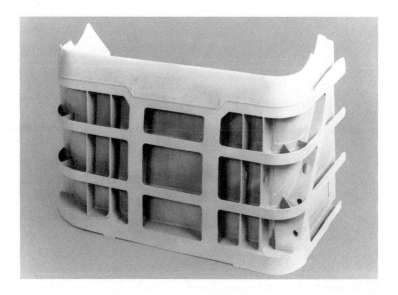

Figure 6 Intake floor (aircraft part): low-pressure sand casting (material: A357).

in the test instructions. More data concerning this example are published in Mietrach (1985). Table 4 verifies that the original target data of cost savings have been exceeded very notably.

■ Other examples

There are other examples from Europe and the USA which show the increased use of castings in aircraft construction to reduce production costs.

Only the following applications will be given here because others would go beyond the intent of this paper:

- Krüger-Flap of Boeing 737 (size: 1950 × 360 × 2 mm).
- Part of the Airbus (size: 850 × 500 × 300 mm), Figure 7.
- Electronic housings for aircraft (size: 320 × 140 × 160 mm).
- Part of CFM56 engine (size: diameter 1050 mm), Figure 8.
- Airbus part (size: 450 × 25 × 140 mm).
- Helicopter part (size: 150 × 150 × 50 mm).
- Fixture for a military aircraft (size: 240 × 155 × 265 mm), Figure 9.
- Cockpit frame of US fighter (size: 3660 × 1220 × 600 mm), Figure 10.

Table 4 Weight and cost calculation for the MRCA Tornado 'intake floor' précial and low-pressure sand casting.

	Old: Many single parts	New: Casting-version			Weight (%)	Costs (%)
Number of parts	13 machined parts 9 sheet-metal parts c. 400 fasteners	1 casting	Précial casting	Goal Achieved	±0 −1	−10 −18
Weight (kg)	9.60	Précial casting 9.50 Sand casting 9.90	Low-pressure sand casting	Goal Achieved	+10 +3*	−35 −60

*For minimum wall thickness $s = 2.5$ mm, today with low-pressure sand casting $s = 1.8$ mm possible.

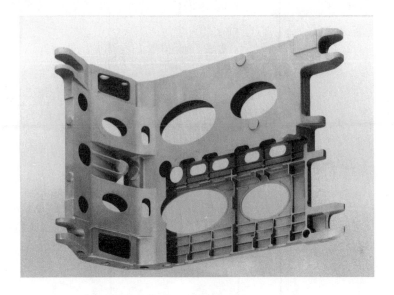

Figure 7 Airbus part:low-pressure sand casting (material: A357).

Figure 8 Part of CFM56 engine investment casting (material: 17.4 PH; size: diameter 1050 mm; weight: 70 kg).

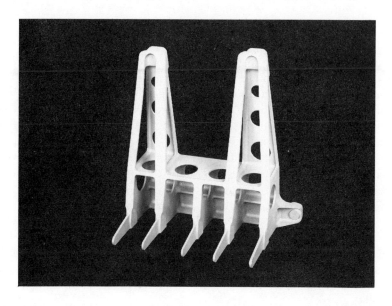

Figure 9 Fixture for military aircraft investment casting (material: A357; size: 240 × 155 × 265 mm).

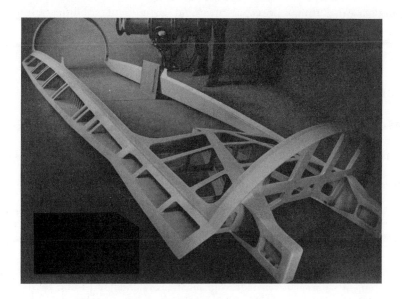

Figure 10 Cockpit frame (US aircraft) sand casting (material: A357; size: 3660 × 1220 × 600 mm; thickness: 3.5 mm).

■ Summary and prospects

The drastic increase in the cost of raw materials, energy and labour in recent years has inevitably brought about higher structural costs in airframe production. The development of economic production processes such as the advanced aluminium casting technology promises to be a remedy for this situation. The aluminium casting technology is an extremely economical production process by which the structural costs – and the weight – can be reduced drastically. It is clear that the advantages of casting only have an effect if the component design is adjusted to the production process (castings-related design). To achieve this, the designer and caster must co-operate closely from the beginning of component development.

Examples from the USA and some European countries in which castings have been introduced more extensively in aircraft programs have shown this course of action to be successful. Genuine components (NIB, intake floor) from the MRCA Tornado have proven not only the technical feasibility and reliability but also the economy of the aluminium casting technology.

The foundries are pressing on with the development of new technologies (larger components, thinner wall thicknesses, closer tolerances, higher reproducibility) and materials (better castability, higher mechanical values) in order to increase the share of castings in future structures considerably. Amongst other things, low-pressure sand casting and 'solidification under pressure' have had a positive influence on the strength values.

References

Faber J. W. (1980) *Cast Aluminum Structures Technology*. Summary Technical Report, Technical Report AFWAL-TR-80-3020

Mietrach D. (1981). Lösungsansätze für den wirtschaftlichen Leichtbau von Flugzeugstrukturen. *ALUMINIUM*, Heft 6 und Heft 7

Mietrach D. (1982). Advanced casting – today and tomorrow. In *AGARD*, Brussels, Belgium, 4–9 April 1982

Mietrach D. (1983). Erfolgreiche Einführung wirtschaftlicher und leichter Aluminium-Gußstrukturen im Flugzeugbau: Zusammenwirken zwischen Konstrukteur und Gießer; Ontwerpen en gieten van hoogwaardige constructiedelen. NVvGT/KIVI leergang, Delft, Oktober 1983

Mietrach D. (1985). Nachweis der Vorzüge fortschrittlicher Gußtechnologien an einem repräsentativen Flugzeugbauteil aus Aluminium, I und II. *ALUMINIUM*, **61**(6), S.421/425 und 7, S.510/514

Mietrach D. and Weilke J. (1982). Erfolgreiche Einführung wirtschaftlicher und leichter Strukturen aus Aluminiumguß im Flugzeugbau. *ALUMINIUM* **58**(2), S.97/100 und (3), S. 157/162

2.5
Design for manufacture – a case study

H. Gill
The Polytechnic, Huddersfield, UK

■ Introduction

The Corfield Report (1978) addresses the issue of product design and is generally critical of the lack of investment in design by UK manufacturing industry. The report also draws attention to poor control methods employed when design projects are, however seldom, initiated.

But perhaps the report's most telling comment is in the foreword where it quotes the American Management Consultant, Peter Drucker. After visiting a number of UK manufacturing plants Peter Drucker had remarked that he saw '. . . a lot of people trying to do better that which should not be done at all.'

Developing this general theme the Corfield Report assigns total responsibility for product design to management and asserts that: '. . . The main elements of cost and desirability in the market place are determined, not on the shop floor, but in the design. Poor design wastes valuable labour and materials, impoverishes our industrial output and insults [my underlining] both the customer and the work people who invest their lives in the manufacturing industry. Good design adds value to scarce material, gives satisfaction to the producer and prestige to the country.'

Despite this truth, which for many people appears to be immutable, that design determines most and affects all those factors that contribute to a product's success, there is still abundant evidence that the message is not understood or is disregarded. In the UK, government and other agencies are investing very large sums of money in efforts to improve productivity and/or efficiency of manufacture. Not only is there an absence of parallel investment

This paper is reprinted by permission of The American Society of Mechanical Engineers from *Proceedings of the International Conference on Engineering Design*, held in Boston, MA, August 1987 in design but there seems to be a limited, if any, effort to ensure that the products to which this improved manufacturing efficiency is directed are well designed or are the 'right' product in the first place. The most efficient production imaginable adds little to the wrong product.

The Corfield Report again: 'Since 1950 the UK has consistently lost market shares both within the UK itself, where imports have been taking an increasing share of the home market, and in the world market. Other industrial economies have managed to hold on to their shares in the world market, for example, France, while some, for example, Germany and Japan, have dramatically increased theirs.'

The case study that is the subject of this paper illustrates, in the author's view dramatically, what can be achieved by first getting the design right.

■   A case study

☐   The company: historical background

The company was founded in 1904 by Ernest Brook who began his business manufacturing a 1-HP single-phase electrical machine. This was quickly followed by three-phase machines, and with the emphasis on volume production on a relatively small range of motors, they have, under the name Brook Motors, specialized in the manufacture of induction motors, mostly of the squirrel cage design.

In 1972 the company was taken over by the Hawker Siddeley Group, a multi-national company with an annual turnover of almost £1600 m. Hawker Siddeley have wholly owned subsidiaries in 50 countries and operate 130 companies, in five major product groupings. They had, in 1968, taken over the Crompton Parkinson Company who had expanded by acquisition from beginnings in the late 1800s. Part of the Crompton Parkinson product range was electrical machines and Hawker Siddeley rationalized the product ranges of both Brook Motors and Crompton Parkinson and formed a new Company: Brook Crompton Parkinson (BCP).

It is this name that the company now trades under and this study is based on initiatives at what was formerly Brook Motors, Huddersfield. BCP have a turnover of around £70 m and employ 2000 people; the Brook Motors component, having a turnover of £40 m and with 1100 employees, is just over half the business. More than 25% of output is exported directly with indirect exports, that is, motors exported as part of customers' machinery, accounting for additional exports. The company is then, clearly, an important part of the local economy and is making a useful contribution to the UK economy as a whole.

☐ The product

The study is based on a range of squirrel cage induction motors. A typical example is shown at Figure 1. The overall design is traditional, that is, a wound stator is fixed to an outer case known as the yoke. This stator interacts with a shaft-mounted rotor and rotary motion is 'induced'. Endshields carry bearings for supporting the shaft and serve also to enclose the whole assembly. The motor is cooled by a shaft-mounted fan which is, in turn, protected by a fan cover. A number of design changes were made but the most significant of these centred on the yoke and it is this component that this study is mostly concerned with.

☐ The problem(s)

The product is required in a bewildering number of variants. These can be due to:

- the needs of different standards;
- the needs/preferences of different countries;
- foot, flange or pad mounting;
- different positions of terminal box;

and there are others.

There are three variants resulting simply from preference as to the position of the terminal box: for example, when viewed on the output shaft, the UK and Germany prefer terminal box 'right', Canada and the USA prefer terminal box 'left' and terminal box 'top' is the preferred arrangement in France and Sweden.

In the old design the yoke was gravity die cast in aluminium. From the foregoing it is clear that yoke die design and yoke production presented many problems: yokes were cast with or without feet, with different terminal box positions, in different lengths to comply with standards and so on. Yokes with integral feet caused problems due to differential cooling of what is clearly an asymmetrical shape. Dies were in four segments forming the ribbed periphery and foot profile, a central mandrel formed the yoke bore, and top and bottom plates enclosed the die. This complexity severely reduced die life.

And so a picture emerges of a very large number of expensive dies with a limited life and an ever growing inventory of stocked parts and assembled motors to deal with the variety of demands from the market. All this made the necessary quick response to orders for 'stock' motors and the meeting of ever shorter delivery times for 'specials' very difficult and very expensive to deal with. A further problem with the old design was that machining of feet,

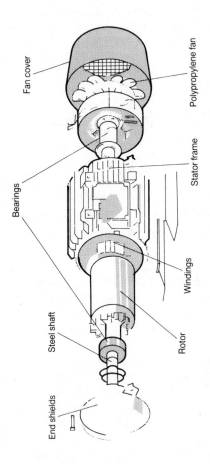

Figure 1 Squirrel cage induction motor.

Fan cover

Polypropylene fan

Bearings

Stator frame

Steel shaft

Windings

End shields

Rotor

86

machining to remove the casting taper of the bore, and drilling and tapping for various fixings, added considerably to the cost. Add to these difficulties an increasingly competitive market and it is easy to understand that the company found their very viability threatened.

☐ A design initiative

The company had always pursued a policy of continuous product improvement and, at first, there was no clear perception of what design changes seemed to be required. Changing the yoke design was given priority since so many of the problems centred on this component.

Motors are designated primarily by 'frame size' and the ranges that form the basis of this study are 63 to 180. There are many variants within this range but the basic range covers 14 frame sizes. Some frame sizes were supplied in two lengths, for example, D180M and D180L:

The figure '180' = the distance in mm from the base of the foot to the output shaft centre line and is the designatory dimension.

The suffixes M and L = medium and long, respectively.

This means that for a given 'nominal' frame size a higher rated (more powerful) motor could be supplied but this usually required a longer stator pack and thus a longer yoke. The distance between feet-fixing holes was also longer.

Rationalization was a clear design need and two strategies were pursued. The first strategy provided for only minimal changes (made possible by improved electrical design) but the feet were still integrally cast and thus the design aim of reversibility was precluded since the feet were not symmetrical about the yoke mid-line.

The second strategy pursued the idea of detachable feet and this paved the way to dramatic improvements.

☐ The new design

Improvements in a computer package developed within the company facilitated much easier simulation of the variables that determine motor performance. This led to a stator pack design that could provide for the power output that had previously required a longer yoke: thus, for example, the 180M and 180L motors could now be constructed from the same yoke, the shorter of the two.

Many changes were then effected in the yoke design itself.

First it should be stated that the whole initiative was *design led*, but throughout there was consultation with other departments, particularly with marketing and with manufacture. Consultations with the latter, and especially 'in-house' tool design and tool making, confirmed that *pressure* die casting – as opposed to gravity die casting as was the practice – could provide for many of the features that design was seeking: the pressure casting process could, for example, effectively eliminate almost all the requirements for machining.

Thus in the new design of yoke – now pressure cast – the following features were facilitated, mostly due to the greater accuracy achievable by this process.

- The yoke was now symmetrical and thus terminal box pads could be at either end relative to foot mountings.

- The feet were detachable and could thus be fixed in any radial position relative to terminal box pads.

- Drilling and tapping for feet fixings, terminal box fixings, fan cover fixings and rod-mounting pad fixings was unnecessary since holes were now cast accurately enough for 'thread-forming' screws to be used directly in the cored holes.

- The fan cover could be located on the periphery of the radial cooling ribs without the need for machining.

- The yoke bore was 'cast to size' and parallel with the foot-mounting pads. Thus machining is now restricted only to the end facings and to the recess for locating endshield spigots.

- Shaped 'pockets' were cast into the yoke periphery to provide for a 'captive' hexagon nut for endshield fixing screws, again eliminating the need for drilling and tapping.

Detachable feet for electrical motors were not new; they have also been a feature of industrial gear boxes for many years. However, the feet always had machined locating faces. In this case, such was the accuracy of the yoke casting and of the feet themselves (made from a steel pressing) that machining of either the foot or of its mounting faces was unnecessary. In fact the foot pressing maintained parallelism between the mounting base and the fixing face of 0.05 mm.

But a more important feature was that the design provided for optional mounting in each quadrant: this was unique in the electric motor industry.

The foot pressing design was finalized after consultations with tool makers and materials specialists. It had to provide for:

- easy forming without loss of accuracy;
- the parallelism referred to above had to be maintained;
- fatigue cracks due to strain age hardening had to be minimized;

- a single blank had to provide for both metric and imperial standard feet;
- zinc coating to prevent electrolytic corrosion between steel feet and aluminium yoke;
- blanked foot mounting holes for different fixing lengths. Even though the standard would allow for higher-rated motors in a shorter yoke (see earlier) it still required the different foot mounting hole positions.

Consultations with manufacturers of fasteners resulted in a recommendation to use 'tri–lobular' Taptite Corflex high-tensile screws. These screws have a high constant torque retention thread, a combined hexagon head and washer face – essential for providing the required clamping area – and eliminate the need for anti-vibration locking devices. The screws were zinc plated to prevent electrolytic corrosion.

☐ Customer response

These technical changes were not carried through without regard to customer response; a simultaneous activity was a market survey. Motor stockists are a very important customer for motor manufacturers. Stockists do, of course, have to respond quickly to orders and since the variety of requirements is almost the same for the stockist's customers as for the manufacturer's then a design concept that facilitated rapid 'stock conversion' was very attractive. The rationalization of the range was also an important advantage since the stock levels that had to be held were reduced.

Other customers were equally enthusiastic. The new design had two further 'spin-off' advantages:

- The greater accuracy of pressure casting effected a material saving of 20% and thus lighter weight.
- The quality of finish adds significantly to an improved aesthetic. The new design has an easily discernible better quality finish and generally improved appearance.

☐ Results

The company is justifiably pleased with the results of this *design-led* initiative; the benefits are many and truly dramatic. The rationalization due to improved electrical design, but more important, carefully considered 'design for manufacture', has reduced the number of yoke combinations to cover the range of motors discussed from over 80 down to nine. This calculation is based on the possible variations in the yoke casting for *one* frame size to cover:

(1)   Terminal box right.
(2)   Terminal box left.
(3)   Terminal box top.
(4)   No feet (for example, flange mounting).
(5)   Pad mounting (for air-conditioning duct applications).
(6)   Metric feet.
(7)   Imperial feet.

A saving of 20% in material cost – with consequent weight reduction – due to improved accuracy allowing reduced section thickness is another major achievement. The reduced number of dies required has saved £1 m in capital outlay and the reduction in variety has reduced stock inventories by £50 000 – excluding interest charges.

The cost of purchasing and fitting detachable feet is more than offset by drastically fewer machining requirements – with attendant reduction in set-up times – longer production runs, more economical batch sizes and less handling in-between machine operations.

Benefits to customers, especially stockists who can now easily convert stock holdings to suit varied requirements, are significant: the company has increased its market share in this range of electrical machines by 8%. Productivity has increased by three to four times and the saving on the cost of a yoke is 30%, leading to an overall saving on the works cost of a motor of 8%. This, at a time of economic recession and increasingly severe competition, has maintained, indeed improved, the company's viability.

☐   **Conclusions**

Much has been said and written about the benefits that should accrue if adequate resources are invested in design and if companies embark on design initiatives such as described herein. The Feilden Report (1963) drew attention to the need to improve arrangements for engineering design, in both industry and education; other reports have submitted the same arguments and even the government has added its voice to the debate. But examples, such as discussed in this paper, of companies designing their way out of difficulties and consolidating their competitive position, are too rare an occurrence to sustain any lasting optimism.

The Feilden Report also states that '. . . engineering goods are sold on the merits of their performance, reliability, appearance, delivery and price . . .' and that '. . . design determines most and affects all of these factors.' Too much emphasis has been placed on first cost despite the evidence in Rothwell *et al.* (1983) and many other publications that non-cost factors are often as important, if not more so. Now, the UK, notwithstanding a devaluation of the pound against most other major currencies, is, for the first time in its history, a net importer of manufactured goods.

Self-sufficiency in energy, North Sea oil and gas has masked what would otherwise be a major balance of payments crisis. Whole industries have declined and/or disappeared altogether. Tourism and other 'invisibles' fall orders of magnitude short of providing for the hole left in the economy of the UK due to the continuing decline of much of its manufacturing base. Estimates indicate that by the end of the 1990s the UK will once again become a net importer of energy. Time is short. This case study illustrates what can be done. Industry can take a lead from Brook Crompton Parkinson.

References

Corfield K. (1978) *Product Design*. NEDO
Feilden G. B. R. (1963) *Engineering Design*. HMSO
Rothwell *et al.* (1983). *Design and the Economy*. The Design Council

Acknowledgements

The author wishes to express his thanks to Brook Crompton Parkinson and in particular their Director of Engineering, Peter Greenwood, for their help and encouragement in writing this paper.

2.6
Integrating product and process design

Gina Goldstein
Associate Editor, *Mechanical Engineering*

Not long ago, analytical instruments such as spectrophotometers, liquid chromatographs and thermal analysers could be expected to last as long as 10 years. But today, with new competitors continually entering a market that is itself in constant change as a result of technological progress, the useful lifetime of these instruments rarely exceeds three to five years. In such an environment, a company that can come up with new products quickly and at a comparatively low cost has a clear competitive edge.

■ Global improvement plan

In 1984, Perkin–Elmer Corp, the world's largest supplier of products and services for chemical analysis, instituted a global quality-improvement plan extending across both product and process lines at its Connecticut Operations Sector in Norwalk. Inventory turnovers were to be increased by reducing set-ups, lot sizes and parts, and by improving scheduling policies and communications with vendors; manufacturing operations were to be consolidated to improve work flow and in-house lead times. Internal and external product failures were to be reduced by a factor of 10 over the course of five years. In product development, the period from completion of the first prototype to market was to be reduced to six months for new models or accessories and 12 months for new products. Teams of representatives from manufacturing, marketing, quality assurance, service and engineering would work together to achieve this goal; design for assembly and design for manufacturability techniques would be used to create more integrated

This paper is reprinted by permission of The American Society of Mechanical Engineers from *Mechanical Engineering*, April 1989, pp. 48–50 products, each of which was to contain 40% fewer parts than its predecessor. Finally, CAD and manufacturing systems would be integrated and design rules standardized so that products could be designed at one site and built at another or manufactured at several sites concurrently.

□   **Consolidated manufacturing**

Since electronics are used in all of Perkin–Elmer's instruments, a pilot project in just-in-time manufacturing was established for the assembly of printed circuits. As Joseph F. Malandrakis, the company's vice-president of operations, recalls: 'When we first looked at the master plan, we had two totally different manufacturing operations – an instrument group and a semiconductor group. The manufacturing manager of the semiconductor side and the instrument manager sat down and agreed that it really didn't make sense for us to have a facility here and a duplicate facility a mile and a half up the road. It's a common process; let's share the technology.' Accordingly, the entire printed circuit board operation for instrument and semiconductor products was consolidated with the instrument-manufacturing plant.

The specific goal in printed circuit assembly was to improve inventory turns by reducing set-ups, increasing productivity and standardizing parts. New automatic insertion equipment has reduced set-up times by 84%, allowing the department to reduce lot sizes and co-ordinate them with the number of instruments actually being built on the other side of the floor.

'Before we had this equipment,' Malandrakis says, 'it probably took 30 minutes to an hour to set up for a job. Now it takes 10 minutes. With the old equipment, we might have built 100 boards or so, and 80 of them would have gone into stock to be used up over a few weeks, or even months. Now we're building boards in lot sizes of about 20.'

'We did an analysis to determine which parts are used on 80% of the boards,' Malandrakis continues, 'and they're loaded on that equipment full-time. The other 20% are on or off depending on the set-up of the job. But our goal is to load up just once and never change it for any product.' Since large buffer, work-in-process, and raw material inventories tend to obscure problems, these reductions have greatly improved efficiency.

The new equipment also enabled an entire initial inspection operation to be eliminated, resulting in significantly fewer failures from handling. In just three years, Perkin–Elmer has seen a 100% improvement in reducing circuit board failures the first time through the test equipment. The goal this year is to realize another 100% improvement.

With the initial inspection operation no longer necessary, the next step is to expedite assembly even further by means of vendor co-operation. Malandrakis explains: 'Through statistical process controls, I want to be able to go straight from my vendors, right on to the equipment. And that's probably from six months to a year down the road.' Previously non-standardized parts like the power supply are being standardized, and a new vendor quality-rating system is helping the company to evaluate its suppliers as to their quality and delivery. Other goals are to greatly reduce the number of vendors from the current level of over 1000 and to work with them more closely.

All together, these improvements have greatly increased yields. 'In just two years,' Malandrakis says, 'the lead time for printed circuit boards has been reduced by 60%.'

When it comes to manufacturing the instruments themselves, many of the same goals and strategies were implemented. All the instrument division's products are built in a single assembly area. 'It's hard to envision,' Malandrakis observes, 'but last year we had 40 000 more square feet devoted to building instrumentation. What we have now is all the process groups circling the final assembly and test area, feeding it parts. We have it set up so the instruments flow from assembly to the test cells to shipping. You don't need cluttered workstations, which actually tend to accumulate inventory and cause inventory and technical problems. Now it's all out in the open.'

As the various instruments queue up, workers in assembly and test pace the line themselves. 'They can see what's coming in,' Malandrakis says, 'and they have the flexibility to move to another instrument or accessory as needed.' Another benefit of consolidation is that communications have improved and led to greater efficiency. 'Since the test people are so close to the assembly people, they can walk over and say, "Hey look; you forgot a lockwasher here." And the assembly person can fix it and check the rest.' Moreover, if rework is required, only five or six instruments tend to be involved because the set-ups are kept down and only a few units are on the line.

Previously, it would take two or three days to assemble one of the company's typical IR products, the FT-IR-1600. Today, the FT-IR-1600 can be assembled and tested in roughly eight hours. One reason for this is self-diagnostics. 'We do some manual diagnostics,' Malandrakis explains, 'but there are computer programs that allow these units to test and diagnose themselves. And the software was written by a manufacturing engineering software group as well as development engineers.' This high level of automation, coupled with a highly manufacturable design, facilitates training. One technician in each of four test cells can test four instruments at the same time.

Another change is in part numbers. Six years ago, the IR spectrophotometer would have had four times more components than it does today. As a result of this drastic reduction, according to Malandrakis, the instrument 'is a pleasure to build, test and ship out. Now we're finding out that in the vast majority of our installations, the customer has this product up and running even before the service engineer gets there. And that's setting the standard for new products.'

■ Teamwork

In its new products and models, Perkin–Elmer is striving to design instruments not only with fewer parts, but also with fewer configurations. To meet these goals, product development teams composed of representatives from quality control, engineering, manufacturing, purchasing, marketing, sales and service participate from the earliest stages. The FT–IR–1600 was the first instrument built with the help of one of these product development teams under a new set of development guidelines.

The team concept has also been applied to quality improvement. 'Problem-solving teams formed directly from the labour base are identifying problems in their own areas and fixing them. These are the types of problems I would never have enough time to look into,' says Malandrakis, 'but they know where the problems are because they work there every day and they can fix them just like that.'

So far, eight teams have been formed to devise comprehensive strategies for improving products and processes, such as the infrared and atomic absorption spectrophotometers, and processes, such as PC board assembly. Another process area that has participated in the improvement plan is the sheet-metal operation, where the use of a computer-aided process planning system, CAPP, is now standard in determining fabrication requirements. Brigham Young University's group technology software program is used to calculate material uses, operation sequences, finish requirements and standard labour hours. This information is then used to create and route parts automatically, create bills of material, and create and route NC tapes.

■ Consistent plans

The manufacturing engineer, prompted by the program, supplies data directly from the part drawing to establish a part database. The process plan generated includes all the paperwork required to introduce the part order into the MRP system for subsequent fabrication. At the moment these forms are processed through the keypunch group for implementation into the manufacturing system, but eventually that step will be eliminated and the CAPP data will be electronically transferred directly from the group technology software to the MRP system.

'Before the CAPP system was brought in, if you asked three different manufacturing engineers to provide a process plan for a particular part, chances are you'd be given three different process plans', Jack Herman, manager of the fabrication manufacturing engineering group, recalls. Now, the decision-making expertise captured within the group technology software enables engineers to create a single, consistent process plan. The two primary benefits of the system are, first, that the paperwork moves through the system much quicker, and, second, that consistency in the way parts standards are set is ensured. 'If we find we're not quite right,' says Herman, 'we have a common database and can go back and fix it.'

'When we did an evaluation of this system,' Herman continues, 'we knew there was going to be a cost saving, but it turned out to be really fantastic.' A typical sheet-metal part that in the past took an hour to complete can now be produced in 10 minutes. And a weldment, which used to take four hours, today requires only 40 minutes. The CAPP system has also proven cost effective and dependable in identifying the labour and material constituents of total part costs. The database of labour content and rates permits fabrication costs to be estimated at an early stage in the design process. Finally, the drastic reduction in the labour involved in generating part documentation has given engineers the opportunity to spend more time on the shop floor. 'Before, 30% of the engineer's time was devoted to clerical work', Herman says. 'Now that time can be devoted to thinking up new ways of improving processes on the floor.'

Perkin–Elmer is now developing another CAPP system for its machine shop parts fabrication plans. The mill/turn machining centres were chosen as the first module for developing CAPP in this area. The effort will be carried into the various milling machining areas until all the required CAPP modules have been completed within the machine shop operations. The tooling classification for perishable cutting tools will initially be used to identify optimum machine, feed/speed, and type of cutting tool based on part configuration and material. This will pave the way for standardization that will minimize redundant tooling and inventory investments.

In a related program, the fabrication manufacturing engineering department at Perkin–Elmer has applied paperless factory software to the procedure for obtaining documentation for work order releases. Enhancements to the existing part routing program now allow a combined route/op sheet to be created directly on the MRP system's CPU; printouts of part lists are then obtained directly from the MRP system.

Before the paperless enhancements were added, Herman says, five days would pass before all the paperwork was in place on a work order. The process entailed many time-consuming steps: obtaining a route sheet printout from the MRP system, determining if an op sheet was required, pulling a copy of the op sheet from a file, making a list of all the required drawings, making copies of op sheets and refiling them. The on-line routing and operation database has eliminated keypunch operations; hardcopy op sheets; the need to obtain, reproduce and refile op sheets; and the need to review documentation and manually prepare a list of required drawings. As a result, the process now takes three days.

■ Role of CAD/CAM

No improvement plan would be complete without CAD/CAM. Perkin–Elmer is now running an Intergraph CAD/CAM program on two converted DEC VAX machines for tool design and for NC programming of sheet-metal parts. Drafting boards have been completely eliminated in tool design, except when old fixtures have to be updated. As a result, blueprints can now be sent to the tool room for fabrication in about a quarter of the time it took in the past. The library of components – tooling plates, pins and clamps – has also helped in the effort to standardize parts.

The CAD/CAM system has had a similar effect on the NC programming. In addition to eliminating errors of calculation, the system has helped to minimize shop set-ups. The library of standard tooling for sheet-metal parts in the system database has been reduced from 600 to 16. Eventually, it should be possible to make all sheet-metal designs with those 16 tools. A verification routine enables NC programs to be tested on screen so machines on the shop floor do not have to be tied up.

Thus, quality improvement is a principle that is being enforced throughout the company, from design engineering to the shop floor. As Joe Malandrakis says, 'That's the way you're going to succeed in this type of market place. You can't keep doing the same old things over and over; you've got to continually improve your entire process.'

Part 3

Manufacturability Evaluation Methods

■ Introduction

The papers presented in this part generally describe those methods that are aimed at supplying information to designers so that they can assess various design alternatives throughout the design process. Ideally, these methods should be qualitative and quantitative, and should ensure that the designer arrives at the optimum configuration with regard to function and economic manufacture. The papers concentrate on those methods developed to assist the designer to *design for manufacture*, DFM, rather than those that concentrate on understanding the designer's actual thought process. Researchers in this latter area have attempted to develop theories and methodologies that can be applied universally, and in fact many design theorists have claimed to have developed methods that can be applied to all types of design (Yoshikawa, 1989). The fact is, however, that, for a given specification, different people will produce different designs. Franke (1986) claims that in reality methodical design tools are hardly ever used, and this is probably true for the type of methodology that defines the design processes via a standard set of algorithms. Design is far too complex to reasonably expect that this approach could ensure the trade-offs that have to be made by the designer in order to reach a compromise solution after considering what may be many conflicting requirements.

The papers presented here, however, have been selected on the basis that they describe methods that are being used to good effect in industry. None of these methods attempts to replace the design function, rather they are tools that indicate areas of the design or design parameters that can be improved. The fact is that no bell rings automatically when the best solution is discovered, and the designer still relies on judgement, creative talent and experience.

In his book *Engineering Design Methods*, Cross (1989) describes the nature of problems tackled by designers as being 'ill defined' or 'ill structured', in contrast to well-defined or well-structured problems such as chess playing, crossword puzzles or standard calculations. He indicates that ill-defined problems are characterized by the following:

- There is no definitive formulation of the problem.
- Any problem formulation may embody inconsistencies.
- Formulations of the problems are solution dependent.
- Proposing solutions is a means of understanding the problem.
- There is no definitive solution to the problem.

Cross also identifies two types of design methods: creative methods, which are intended to help stimulate creative thinking, and rational methods, which encourage a systematic approach to design.

The vast majority of the methods discussed in this part fall into the latter 'rational' category. Creative methods have received a great deal of attention in the past and there are many publications on the subject. However, there is an almost complete lack of evidence to show that these methods are being used by manufacturing companies. Until recently, little attention has been given to DFM, which requires a systematic and disciplined approach. The potential of improving the design of products by working in a clear methodical manner is immense, although this is no substitute for intellectual effort. In fact, the two are complementary, and the greatest chance of success will be achieved by ensuring the correct balance between the rational and creative aspects of design. The designer should find that the methods described in this part will in fact assist creativity, and careful documentation of his systematic thoughts will leave more time for intuitive and creative thinking.

■ Discussion of papers

In the first paper presented, Stoll gives further consideration to many of the principles and rules covered in Part 2, as well as quantitative evaluation methods and computer-aided techniques, which are covered in Part 4. In addition to discussing empirically derived DFM rules, Stoll comments on the axiomatic method proposed by Suh and Yasuhara (1980). He believes that 'design axioms cannot be proven, but rather must be accepted as general truths because no violation or counter-example has ever been observed'. Suh and Yasuhara have, after originally proposing several axioms, reduced these to the following two fundamental axioms:

(1) In good design the independence of functional requirements is maintained.

(2) Among the designs that satisfy Axiom 1, the best design is the one that has the minimum information content.

A conclusion that can be reached from these axioms is that over–design can result through the specification of more than the necessary functional requirements. This should not, however, detract from the importance of starting the design process with a fully comprehensive specification. Oakley and Pawah (1983) found, during their research into a number of engineering companies, that most did not prepare or work with formal specifications. This would seem to indicate that they are not fully in control of their business.

In his discussion of quantitative evaluation methods, Stoll has concentrated on the Boothroyd and Dewhurst *design for assembly*, DFA, method. DFA methods often supply a list of rules, or in Boothroyd and Dewhurst's case a coding system, which indicates the relative ease or difficulty of feeding, handling, manipulating and assembling products. A comparison is made with an 'ideal', if not practical, design to identify areas that should be considered for redesign. Such methods stimulate creativity by encouraging the designer to improve his design with regard to manufacturability. Stoll reminds us that DFM is a large subject spanning many disciplines and viewpoints, and he believes that we are a long way from understanding the many complex and interrelated issues associated with it. It is worth re-emphasizing the benefits of teamwork in dealing with the complexities and interrelated issues of modern design. Further, a clear systematic approach is required so that the specialists within the team can make their contributions at the right time in the process.

The next paper by Shoemaker and Kacker proposes a method for planning experiments to achieve robust designs that perform reliably as specified and are immune to variations, which are difficult to control. The method is compared to the Taguchi method described by Brown *et al.* later in this part. The four basic steps described by Shoemaker and Kacker are:

(1) Formulate the problem. Once again, team work is emphasized to highlight objectives and list the product variables.

(2) Plan the experiment. The method proposes two stages: a 'control array', to plan for varying the settings of the control parameters, and a 'noise plan', to specify how noise variables will be explicitly varied in the experiment. Noise variables are those that the designer finds difficult or expensive to control – for example, tolerances, material quality, custom environment, wear, and so on.

(3) Identify improved settings of the controllable variables from the results of the experiment.

(4) Confirm the improvement in a small follow-up experiment.

Such detailed experimental methods are obviously not suitable for all products and will not be generally appropriate for low-volume, high-cost products, where the cost of hardware and the time ideally required for multiple experiments can be prohibitive. Also, computer simulation methods – for example, mechanism and finite element analysis – may be more beneficial for machine design applications.

The paper by Brown *et al.* describes the Taguchi experimental method together with other methods that have been tried to good effect within a large engineering company. An interesting feature of this paper is the way various methods are shown to interrelate. A case study is included, which illustrates how the Taguchi method can be used to gain a better understanding of the critical parameters associated with the design of small electromechanical parts. The study illustrates the value of the method in improving designs and reducing development time-scales for future generic products. It also confirms the need for much experimentation. The concept of a total quality organization is introduced with the *quality function deployment* method being proposed as the 'corner stone of sound quality'. Quality function deployment is probably the most complete method, in that it starts with the specification or 'voice of the customer' and proceeds by a series of matrices to evaluate product features, component characteristics, process characteristics and production operations. The method is enhanced by being used in conjunction with other methods, in particular the Taguchi method and DFA, to evaluate various features and characteristics. The Taguchi method is used to good effect in distinguishing between the many mundane parameters and those that are important.

Statistical methods feature strongly in Taguchi's methodology, which utilizes statistically designed experiments to minimize the deviation from target values, caused by factors that are out of the user's or manufacturer's control. A danger from experimental methods can be a tendency towards the 'analysis and paralyse' phenomenon, in which an over-emphasis is put on experimental methods. For example, a product that has 12 variables, each with two values, would require 4096 different experiments in order to investigate all possible situations. An important feature of Taguchi's statistical technique is to reduce, significantly, the number of experiments to achieve the same effect. The fundamental aim is to achieve a 'robust' design at minimum cost. However, even with these methods, experience is needed to maximize the value of decision making with regard to the importance of the interdependence of the different design parameters. Additional methods discussed are *failure modes and effect analysis*, as well as others used within the company to improve manufacturing processes and overall quality. The paper emphasizes the importance of keeping abreast of the advances being made in production technology, as these may make possible those designs that were once thought to be impractical. An example of such a case is the control of a six-axes CNC machining centre directly by a CAD solid model input, which can make possible the description and machining of extremely complex surfaces, such as found on turbine blades. This has been utilized effectively in the past with the overall operating efficiency of the turbine being enhanced.

Probably the best-known and widely used method is the DFA method developed by Boothroyd and Dewhurst, which is discussed by Stoll in the first paper presented in this part. In their paper, Boothroyd and Dewhurst discuss their DFA technique, refer briefly to axiomatic approaches and then go on to describe their latest work, which is aimed at supplying designers with cost estimates early in the design process. There is now much evidence available to demonstrate the tangible benefits that have been obtained through the use of the DFA method. Some of these are indicated in Table 1. A particularly beneficial feature of DFA is that it distinguishes between the design of the product and the design of individual components. It tackles the cost of the product or assembly with regard to manufacture, whereas DFM generally tackles independent features of individual components, not those that interrelate with the other parts in the product. DFA seeks to eliminate

Table 1 Design for assembly achievements.

Company	Product(s)	Reduced parts count (%)	Reduced assembly cost or time (%)
Lucas (UK)	Automotive (8)	Range 18–62	Range 28–65
NCR (UK)	Cash dispenser	60	70
DEC (USA)	Computer mouse	50	72
Motorola (USA)	Battery charger	42	65
Clesse-Mandet (France)	PR valve	61	74

After Miles (1989).

parts, as discussed earlier by Stoll and others, and Boothroyd and Dewhurst claim that the cost savings achieved through the reduction of parts is often far higher than the actual assembly cost reductions. The elimination of parts also has other benefits, but these are more difficult to quantify. Examples of these benefits include cost savings by reductions in component documentation, process planning, quality assurance, production control, storage and distribution.

The paper is critical of the axiomatic approach, claiming that it has weaknesses directly related to cost. Such an approach 'does not provide any means of making judgements between the centrally important trade-offs posed by possible alternative choices of different materials and processes. Second, at the detail level, guidelines tend to lead designers in an essentially fruitless direction. This is because manufacturing guidelines are invariably intended to make individual process steps as efficient as possible.' It follows that this approach will result in concentrating on features of individual components with no attempt being made to make savings in assembly costs.

The commentary for Part 2 noted that the designer must have access to reliable cost data if he is expected to produce cost-optimized designs. Boothroyd and Dewhurst have addressed this problem and are now developing a DFM system, which is described in the latter half of their paper. The aim is to predict both assembly and manufacturing costs during the early stages of design. Software packages have been developed that can be used to obtain costs associated with processes such as injection moulding and machining. These consider the dimensional and quality-related data, together with part and mould complexity, to arrive at a cost breakdown for the component under investigation. Clearly, at an early design stage the cost estimates will have limited accuracy. Also, actual costs will vary from company to company depending on many factors, including their costing structure, production facilities, and so on. However, the software could be tailored to suit individual companies, and the technique supplies the designer with, at the very least, relative costs and an indication of those high-cost areas that require his attention in order to evaluate possible design changes and trade-offs.

An example of the use of the DFA method is given in the paper by ElMaraghy *et al.* This demonstrates how each subassembly and the final assembly was analysed to arrive at a quantifiable design efficiency with which the redesigned product could be compared. The analysis took into account the company's production facilities, development times, tooling and the overall manufacturing cost. Those areas that required the designer's attention, in order to improve design efficiency, are highlighted. In the example, the design efficiency was improved by:

- using sealed bearings to eliminate oil throwers;
- eliminating fixings and fastenings through the use of crimping methods;
- using rigid ground terminals moulded into the connector in place of flexible leads and terminal clips.

This latter improvement meant that there was no need for orientation when the endcap was fitted, so the components could all be assembled with the stacked 'top-down' operation with no hard-to-reach areas. These features, together with the design of self-locating parts, allowed the use of automatic assembly techniques and also benefited manual assembly. As a direct result of using the DFA method, a 68% parts reduction was achieved, which led to a 60% reduction in assembly time. The example shows nicely how a disciplined approach can enhance, but not replace, the designer's creativity and experience.

The final paper by Williams illustrates that worthwhile cost savings can be made by looking at existing products, as well as new product concepts. In reality, the high level of cost savings that are possible with new concepts are unlikely to be achieved because of existing tooling and the need to ensure backward compatibility (for example, retrofittable parts). However, the study shows that, assuming that the life expectancy of the product is sufficiently high, worthwhile cost savings can be achieved. The overall manufacturing costs of the existing products were reduced by between 5 and 19%, although details of how these cost savings were achieved are not given. However, the philosophy and approach are described, and the normal requirements for embarking on the redesign of existing products are indicated. For example, any redesigned project with an estimated completion time of more than 12 months was automatically eliminated. Potential projects for redesign were generally considered in areas where:

- technologies had advanced since the product came onto the market;
- parts could be reduced through redesign;
- a new design approach could alleviate concerns with reliability.

All of the papers presented in this part show how it is possible to enhance the design process by working in a clear methodical manner. While it is recognized that a universally valid system is not applicable, there is no doubt that a suitably structured methodical approach will give a clear indication of the many important points that need to be considered. This does not in itself offer a substitute for concentration and innovative or inventive ability, but it will lead to fewer oversights and encourage the designer to think systematically. The greatest probability of success is ensuring the correct balance between the methodical and creative aspects of design.

References

Cross N. (1989). *Engineering Design Methods*. John Wiley & Sons Ltd

Franke H. J. (1986). Design Methodology and Practice – A Critical Assessment. *KSB Technische Berichte*, 17–23

Miles, B. L. (1989). Design for assembly – a key element within design for manufacture. *Proceedings of the Institution of Mechanical Engineers*, Vol. 203, Part D, *Journal of Automobile Engineering*, 1989

Oakley M. H. and Pawah K. S. (1983). Research the design/production interface: product specifications. *Design Studies*, **4**(1), 13–19

Suh N. P. and Yasuhara M. (1980). A quantitative analysis of design based on axiomatic approach. *Computer Applications in Manufacturing Systems*. ASME Production Engineering Division Publications, PED, Vol. 2

Yoshikawa H. (1989). Design philosophy: the state of the art. *CIRP Keynote Paper*, Trondheim, August 1989

3.1
Design for manufacture: an overview

Henry W. Stoll

Design for Manufacture, Industrial Technology
Institute, Ann Arbor, MI 48106, USA

■ Introduction

A manufacturing system comprises a large number of distinct processes or
stages which, individually and collectively, affect product cost, product
quality and productivity of the overall system. The interactions between
these various facets of a manufacturing system are complex, and decisions
made concerning one aspect have ramifications which extend to the others
(Figure 1). In its broadest sense, design for manufacture, DFM, is concerned
with comprehending these interactions and using this knowledge to optimize
the manufacturing system with respect to cost, quality and productivity.
More specifically, DFM is concerned with understanding how product
design interacts with the other components of the manufacturing system and
in defining product design alternatives which help facilitate 'global'
optimization of the manufacturing system as a whole.

DFM can be divided into a number of subareas. Design for fabrication,
DFF, involves the design of product components and parts in ways which are
compatible with the method of fabrication (that is, design for casting,
sheet-metal forming, injection moulding, etc). DFF has always been an
important part of product design, especially in the mass-production industries. Recent experience with modern, flexible manufacturing and formation
methods, however, has clearly demonstrated that product design and
manufacturing interact in many ways which extend far beyond just part
fabrication. To date, the greatest single opportunity for product design

This paper is reprinted by permission of The American Society of Mechanical
Engineers from *Applied Mechanics Reviews*, Vol. 39, No. 9, September 1986, pp.
1356–64

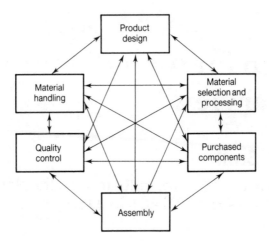

Figure 1 Manufacturing interactions.

improvement using the concept of DFM has been in the area of assembly. This activity has become widely known as design for assembly, DFA, and involves minimizing the number of parts to be assembled as well as designing the parts which remain to be easy to assemble. Other areas of DFM include product design considerations which impact material handling, in-process inspection, quality, etc.

It is also important to note that implementation of the DFM philosophy is strongly tied to the process by which a product is conceived, designed, produced, and eventually brought to the market place to be sold and serviced. Because it is based largely on common sense, the DFM philosophy is, in general, quickly understood and accepted by most product engineers, manufacturing engineers, purchasing agents, managers and even company presidents. The challenge lies in making the DFM philosophy work under the constraint of existing company policy, organizational structure and administrative practice. A significant part of DFM, therefore, involves facilitating change in the way manufacturing businesses are run.

■ Background and organizational issues

DFM is a relatively new way of looking at a very old problem. To appreciate the problem and to understand where DFM is today and what needs to be done in the future, it is necessary to look briefly at traditional practice and at some of the organizational issues which are involved in the DFM approach. The importance of manufacturability in product design has been recognized for years. Just how important is illustrated by the well-known fact that up to

80% or more of production decisions are directly determined by the product design (Lotter, 1984), which leaves very little freedom of choice for process planning. In spite of this, most product design decisions have historically been based on three major factors: product function, product life and component cost (Riley, 1983). Component cost has been, and to a large extent still is, the primary product design consideration involving manufacturability.

Two well-known approaches or techniques have evolved to help the designer deal with component cost. These are value engineering and producibility engineering. Depending on the size and organizational philosophy of the company, these functions may be stand-alone departments, or they may be included as part of either the product design or manufacturing engineering team. Typically, value engineering reports to a manager associated with design engineering while producibility engineering reports to a manager of manufacturing engineering. In general, but not always, the value and/or producibility analysis is performed after product concept decisions have been made.

Obtaining the maximum performance per unit cost is the basic objective of value engineering (Gage, 1967; Mudge, 1971). The value of a product is the ratio of its performance to its cost. In a complicated machine system or product, every component contributes to the cost and the performance of the entire system. The ratio of performance to cost of each component indicates the relative value of individual components. Value analysis, VA, programs, as generally conducted today (Bradyhouse, 1984), first challenge the design of the product searching for simpler designs that will reduce cost while maintaining function. If this effort falls short of the desired cost target, the VA teams move into a secondary effort of challenging the manufacturing methods to identify the needed savings.

Value engineering is primarily concerned with function and its cost. Producibility engineering, on the other hand, is concerned with assuring that parts can be manufactured and assemblies made and tested to meet specifications with available or potentially available techniques, tooling and test equipment at costs compatible with the product's selling price (Howell, 1982). The emphasis in producibility engineering is to protect the interest of the manufacturing function. Presumably, the design engineer seeks to optimize his design to maximize the probability of accomplishing intended functions. Such optimization, if not carefully monitored, could be accomplished at the expense of product manufacturability. In many manufacturing organizations, producibility engineering is seen as the first line of defence, and perhaps the only defence against unreasonably difficult tasks being thrust upon manufacturing (Koenig, 1981).

In the late 1970s, American manufacturers suddenly awoke to the fact that their pre-eminent position in the global market place was being threatened by declining US productivity, a threat from world-class manufacturers, and the emergence of new automation processes. This precipitated a quest for short-term solutions that quickly focused on fledgling technologies

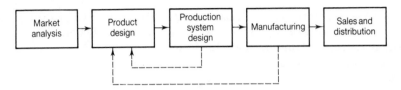

Figure 2   Classical manufacturing model.

such as robotics, flexible manufacturing systems and computer-integrated manufacturing. Failed attempts to implement these and other advanced manufacturing technologies in a variety of industrial settings has taught much about the design/manufacturing interface. Based on these experiences, it is clear to many workers involved with these new technologies that major change in the design process as traditionally practised domestically is needed if promised productivity increases are to be achieved. This realization, born out of often difficult and painfully expensive experience with these new technologies, is an important motivating factor in the development of the DFM philosophy.

The problems creating this need for change can be quickly appreciated by considering the simplified classical manufacturing model shown in Figure 2. In this model, concept decisions, product design and testing are performed prior to manufacturing system design, process planning and production. The experiences with advanced manufacturing technologies referred to above have shown that the design, function and implementation of these technologies are directly related to the product to be produced. Therefore, to properly design a product to be manufactured using advanced, productivity-enhancing technologies, it is essential that interactions between product and process enter into consideration very early in the design process. Product concept and manufacturing concept decisions must be made in parallel in order to obtain an integrated manufacturing system optimally configured to satisfy both product and process needs.

The serial nature of the classical approach prevents integrating design of the product and process, even when manufacturability is recognized as being of paramount importance. For the same reason, value analysis and producibility engineering, although highly valid and worthwhile methods in themselves for reducing cost and improving manufacturability, enter consideration too late in the classical model to offset the fundamental problems inherent in this approach. The result is suboptimal manufacturing system design. Compounding the suboptimality is the needless time, effort and money spent solving manufacturing problems which could have been avoided in the first place through proper product design. Even after the problems are ironed out and the engineering changes made, usually at considerable expense, the advanced manufacturing technologies still fall short of providing hoped-for productivity gains because they are simply not matched to the product being manufactured. Most devastating of all, the extra manufacturing cost and loss of productivity incurred because of this practice continues throughout the life of the product.

The concept of design for manufacture evolved out of this experience and is predicated on the recognition that:

- Design is the first step in product manufacture.

- Every design decision, if not carefully considered, can cost extra manufacturing effort and productivity loss.

- The product design must be carefully matched to advanced flexible manufacturing, assembly, quality control and material-handling technologies in order to fully realize the productivity improvements promised by these technologies.

To maximize the quality of early design decisions and thereby minimize the amount of engineering change, the DFM approach seeks to involve input from as many manufacturing system activities as possible as early as possible. Ideally, convergence to 'globally optimal' product and process decisions should occur at the early, low-cost stages of the project. This approach requires the simultaneous engineering model depicted in Figure 3.

In a study addressing integration of design and manufacturing during deployment of advance manufacturing technology, Ettlie and Reifus (1986) identified five mechanisms which illustrate trends in administrative innovation:

(1)   design–manufacturing team;
(2)   common CAD systems for design and tooling;
(3)   common reporting position for computerization;
(4)   philosophical shift to DFM;
(5)   development and promotion of the engineering generalist.

A major conclusion of the study is that the DFM and engineering generalist adaptations appear to be a necessary part of all successful cases of design–manufacturing integration.

Most domestic companies recognize the need for change and many are trying to implement it. The General Motors Saturn Project and the Chrysler–Liberty Program are two of many well-publicized examples. There have also been some notable product successes associated with the simultaneous (DFM) approach. Among these are the Allen–Bradley motor controller (Knill, 1985) and IBM's redesigned and highly successful new line of typewriters and printers (Pierson, 1985). But change is difficult, and for most domestic companies, real change has not occurred, or is only just beginning to occur.

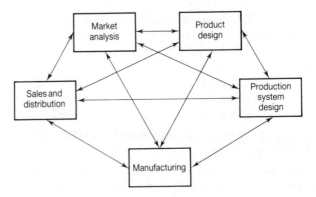

Figure 3   Simultaneous engineering model.

The learning experience associated with implementing advanced manufacturing technology together with the constraints imposed by the classical approach (Figure 2) has caused DFM to develop in many different ways. One approach to implementing DFM is to use an appropriate set of principles and rules to help guide the design of the product and then to evaluate and redesign the product using an appropriate evaluation methodology. This process has been assisted by the developoment of a variety of computer-based and/or computer-aided design programs. Hence, possibly the three main lines of DFM activity at this time involve the development of formalized DFM principles, quantitative evaluation methodologies and computer-aided design programs embodying DFM principles. These activities are directed towards specific needs and types of applications. Organizational structures and administrative procedure are necessary issues when doing any research in DFM. Much of the motivation behind development of the DFM philosophy lies in the need to build company-wide teams which truly work together in the development and manufacture of a product. Promoting and fostering manufacturing team building is therefore a strong underlying theme in most DFM developments as they are being implemented today.

■ DFM principles and rules

To help enable product designers and product engineers to consider manufacturability of the product early in the design process, DFM principles, rules, guidelines, and numerous clever suggestions and tips have been stated in systematic and codified ways. Use of this human-oriented, largely heuristic body of knowledge helps narrow the range of possibilities so that the mass of detail which must be considered is within the capacity of the designer and planner. For the most part, dissemination of this type of knowledge has been through speeches (Schreiber, 1985), trade journal articles (Lewis, 1986), short courses (Altamuro, 1985; Huthwaite, 1985; Stoll, 1986), and in a variety of technical papers (Nevins and Whitney, 1982; Laszcz, 1984; Boothroyd and Dewhurst, 1983a; Suh *et al.*, 1978; Yasuhara and Suh, 1980). There are also some recent books on the subject (Riley, 1983; Andreasen *et al.*, 1983; Tuffentsammer, 1979). In a few cases, both the rules and a suggested methodology for applying the rules are suggested (Lotter, 1984; Bradyhouse, 1984).

Many DFM principles are deeply rooted in the long history of design and manufacture. Most have been learned empirically. Knowledge of these principles and the ability to apply them correctly has always been the hallmark of the experienced expert design and manufacturing engineer. Explicit statements of many of the principles are embodied in both value engineering and producibility engineering. Books which discuss the principles in a variety of formats have appeared on a regular basis over the years (Greenwood, 1959; General Electric, 1970; Chow, 1978). Although the principles themselves are largely invariant and universal, their organization and statement in the context of a DFM philosophy of global manufacturing system optimization give them new value and usefulness.

☐ Product design principles for efficient manufacture

The following DFM principles are discussed to illustrate their global nature and to provide insight into how such principles can be used to aid the product-development team.

Minimize total number of parts

Fewer parts mean less of everything that is needed to manufacture a product. This includes engineering time, drawings and part numbers; production control records and inventory; number of purchase orders, vendors, etc; number of bins, containers, stock locations, buffers, etc; amount of material-handling equipment, containers, number of moves, etc; amount of accounting details and calculations; service parts and catalogues; number of items to inspect and type of inspections required; and amount and complexity of part production equipment, and facilities, assembly and training. Put another way, a part that is eliminated costs nothing to make, assemble, move, handle, orient, store, purchase, clean, inspect, rework, service. It never jams or interferes with automation. It never fails, malfunctions, or needs adjustment.

A part is a good candidate for elimination if there is:

(1)   no need for relative motion;
(2)   no need for subsequent adjustment between parts;
(3)   no need for service or repairability;
(4)   no need for materials to be different.

However, part reduction should not exceed the point of diminishing return where further part elimination adds cost and complexity because the remaining parts are too heavy, or too complicated to make and assemble, or are too unmanageable in other ways.

Perhaps the best way to eliminate parts is to identify a design concept which requires few parts. Integral design, or the combining of two or more parts into one, is another approach (Chow, 1978). Besides the advantages given above, integral design reduces the amount of interfacing information required, and decreases weight and complexity. One-piece structures have no fasteners or joints, and fewer points of stress concentration. Conversely, structural continuity leads to high strength and light weight.

Plastic is a major key to integral design. Plastic is available for making springs, bearings, cam and gears, fasteners, hinges and optical elements. Powder metallurgy, P/M, is a good alternative if plastic parts do not have adequate strength, heat resistance or cannot be held to the tolerance needed (ASCO, 1985). Brazed, welded or staked assemblies of stampings and/or machined parts can often be made as one-piece P/M parts. Extrusions and precision castings (*Precision Metal*, 1975) as well as impacts (Church, 1981) are also good ways to eliminate subassemblies. Although switching to a different manufacturing process may lead to a more costly part, experience with part integration has shown that a more costly part often turns out to be more economical when assembly costs are considered (*Precision Metal*, 1975).

Develop a modular design

A module is a self-contained component with a standard interface to other components of a system (Chow, 1978). Modular design offers the ability to standardize diversity because it allows a product to be customized by using different combinations of standard components. Modular design resists obsolescence and shortens the redesign cycle. A new generation product can realize most of the old modules. Change is provided via a few, new, or improved modules. Cost and ease of service and repair are enhanced because a defective module can be quickly replaced by a good one. Most importantly, modular design simplifies final assembly because there are fewer parts to assemble and each module can be fully checked prior to installation. On the downside, modular design can add cost and complexity because of extra fittings and interconnections required. Therefore, modular design should not be used unless its advantages are needed.

Use standard components

A stock item is always less expensive than a custom-made item. Standard components require little or no lead time and are more reliable because characteristics and weaknesses are well known. They can be ordered in any quantity at any time. They are usually easier to repair and replacements are easier to find. Use of standardized components puts the burden on the supplier and makes the supplier do more.

Design parts to be multi-functional

Combine function wherever possible. For example, design a part to act both as a spring and a structural member, or to act both as an electrical conductor and a structural member. An electronic chassis can be made to act as an electrical ground, a heat sink and a structural member. Less obvious combinations of functions might involve guiding, aligning and/or self-fixturing features to a part to aid in assembly, or providing a reflective surface or recognizable feature to facilitate vision inspection. These latter examples illustrate inclusion of functions which are only needed during manufacture. Such function combinations are often the result of DFM awareness.

Design parts for multi-use

Many parts can be designed for multi-use. For example, the same mounting plate can be designed to mount a variety of components. Similarly, a spacer can also serve as an axle, lever, standoff, etc. Key to multi-use part design is identification of part candidates (Stoll, 1986). One approach involves sorting all parts (or a statistical sample) manufactured or purchased by the company into two groups consisting of:

(1)   parts which are unique to a particular product or model (that is, crankshafts, housings, etc.);
(2)   parts which are generally needed in all products and/or models (shafts, flanges, bushings, spacers, gears, levers, etc.).

Each group is then divided into categories of similar parts (part families). Multi-use parts are then created by standardizing similar parts. In standardizing, the designer should sequentially seek to:

(1)   minimize the number of part categories;
(2)   minimize the number of variations within each category;
(3)   minimize the number of design features within each variation.

Once developed, the family of standard parts should be used wherever possible in existing products and used exclusively in new product designs. Also, manufacturing processes and tooling based on a composite part containing all design features found in a particular part family should be developed. Individual parts can then be obtained by skipping some steps and features in the manufacturing process (Chow, 1978).

Design parts for ease of fabrication

This principle requires that individual parts be designed using the least costly material that just satisfies functional requirements (including style and appearance) and such that both material waste and cycle time are minimized. This in turn requires that the most suitable fabrication process available be used to make each part and that the part be properly designed for the chosen process. Use of near–net shape processes are preferred whenever possible. Likewise, secondary processing (finish machining, painting, etc.) should be avoided whenever possible. Secondary processing can be avoided by specifying tolerances and surface finish carefully and then selecting primary processes (precision casting, P/M, etc.) which meet requirements. Also, material alternatives which avoid painting, plating, buffing, etc., should be considered. This principle is based upon the recognition that higher material and/or unit process cost can be accepted if it leads to lower overall production cost.

Avoid separate fasteners

In automation applications, separate fasteners are difficult to feed, tend to jam, require monitoring for presence and torque, and require costly fixturing, parts feeders and extra stations. Even in manual assembly, the cost of driving a screw can be six to 10 times the cost of the screw (Lewis, 1986). One of the easiest things to do is eliminate fasteners in assembly by using tabs or snap–fits (Chow, 1978). If fasteners must be used, cost as well as quality risks can be significantly reduced by minimizing the number, size and variations used, and by using standard fasteners whenever possible. Screws that are too short or too long, separate washers, tapped holes, and round and flat heads (not good for vacuum pick–up) should be avoided. Conversely, captured washers should be used for reduced part placement risk and improved blow feeding. Self–tapping/forming/locking fasteners are preferred as are screws with dog or cone (chamfered) point for improved placement success. Also, screw heads designed to reduce 'cam–out' problems, bit wear and fastener damage should be used. For vacuum pick–up, use screw heads having flat vertical sides.

Minimize assembly directions

All parts should be assembled from one direction. Extra directions mean wasted time and motion as well as more transfer stations, inspection stations and fixture nests. This in turn leads to increased cost and increased wear and tear on equipment due to added weight and inertia load, and increased reliability and quality risks. The best possible assembly is when all parts are added in a top–down fashion to create a z-axis stack. Multi–motion insertion should be avoided. Ideally, the product should resemble a z-axis 'club sandwich' (Altamuro, 1985; Laszcz, 1984) with all parts positively located as they are added.

Maximize compliance

Because parts are not always identical and perfectly made, misalignment and tolerance stack–up can produce excessive assembly force, leading to sporadic automation failures and/or product unreliability. Major factors affecting rigid part mating include part geometry (accuracy, consistency), stiffness of assembly tool, stiffness of jigs and fixtures holding the parts, and friction between parts. To guard against this, compliance must be built into both the product and production process. Methods for providing compliance include highly accurate (consistent) parts, use of 'worn-in' production equipment, remote centre compliance, selective compliance in assembly tool (SCARA Robot), tactile sensing, vision systems, designed-in compliance features and external effects. Although a variety of combinations of these approaches are commonly used, experience has shown that the simplest solution consists of a combination of acceptable (consistent) quality parts, designed-in compliance features, accurate (rigid) base components and selective compliance in the assembly tool (SCARA Robot: Huthwaite, 1985).

Designed-in compliance features include the use of generous tapers or chamfers for easy insertion, use of leads and other guiding features, and use of generous radii where possible. A clever trick, if possible, is to design one of the product components, perhaps the largest, to act both as the part base (part to which other parts are added) and as the assembly fixture (avoid need for a special fixture to hold the assembly). In any case, the part base should be made as stable and rigid as possible to improve accurate insertion and simplify handling. If fixturing is required, 'fixture-friendly' features such as accurate location points, generous tapers, and other guiding features which provide easy compliance between base part and fixture should be provided.

Gravity is an extremely useful external effect which not only assists compliance, but also costs nothing. In addition to assisting with insertion, gravity is useful for feeding parts and for ejecting finished and defective product. Another external effect is the use of vibration to assist in part insertion (Altamuro, 1985). The vibration insertion process, VIP, consists of nominally locating the part using a robot, vibrating the base until the part lines up and then inserting the part upon alignment. This technique appears to be especially applicable in electronic assembly.

Minimize handling

Position is the sum of location (x, y, z) and orientation (α, β, γ). Position costs money. Therefore, parts should be designed to make position easy to achieve and the production process should maintain position once it is achieved. The number of orientations required during production equates with increased equipment expense, greater quality risk, slower feed rates and slower cycle times. To assist in orientation, parts should be made as symmetrical as possible. If polarity is important, then an existing asymmetry should be accentuated, or a very obvious asymmetry should be designed in, or a clear identifying mark provided. Orientation can also be assisted by designing in features which help guide and locate parts in the proper position. Parts should also be designed to avoid tangling, nesting and shingling in vibratory part feeders.

To facilitate robotic part handling, provide a large, flat, smooth top surface for vacuum pick-up, or provide an inner hole for spearing, or provide a cylindrical surface or other feature of sufficient length for gripper pick-up. Since parts usually come off the production line properly oriented, this orientation should be preserved by using magazines, tube feeders, part strips, etc. Palletized trays and kitting are methods for supplying properly oriented parts to the assembly line. For ease of handling, avoid flexible components. Use rigid gaskets where possible; use a connector to eliminate lead wires; use a circuit board in place of a cable; etc.

Design in features which facilitate product and component packaging. Use standard outer package dimensions for machine feeding and storing, design packaging adequately to protect and ensure quality at all stages of handling, and design packaging for easy handling. Consider material flows within the production facility including product flow, workspace flow, supply flow, hardware flow, trash or scrap flow, bulk material flow, container flow and fixture flow. For each flow, consider how the product, subassembly, component or part can be designed to simplify or eliminate the flow (Stoll, 1986).

☐ **Axiomatic design**

The DFM principles discussed earlier are empirically derived and relate to a variety of specific design situations. N. P. Suh and his associates at MIT have proposed an alternative approach called an 'axiomatic approach' (Suh *et al.*, 1978). In this approach, a small set of global principles, or axioms, are hypothesized. These axioms constitute guidelines or decision rules which can be applied to decisions made throughout the synthesis of a manufacturing system and, if correctly followed, lead to decisions which maximize the productivity of the total manufacturing system, in all cases. By definition, an axiom must be applicable to the full range of manufacturing decisions. Design axioms cannot be proven, but rather must be accepted as general truths because no violation or counter-example has ever been observed. Although several axioms were originally proposed, these have been reduced to the following fundamental axioms as stated by Yasuhara and Suh (1980):

• Axiom 1: In good design the independence of functional requirements is maintained.

- Axiom 2: Among the designs that satisfy Axiom 1 the best design is the one that has the minimum information content.

These two axioms imply that specification of more functional requirements than necessary results in over-design whereas specification of insufficient functional requirements results in unacceptable solutions. Also, as discussed in Yasuhara and Suh (1980), independence of functional requirements must be evaluated within specified tolerances.

Design corollaries are immediate or easily drawn consequences of the design axioms. In contrast to the design axioms, corollaries may pertain to the entire manufacturing system, or may concern only a part of the manufacturing system. Some of the important corollaries given by Yasuhara and Suh (1980) are:

(1) Decouple or separate parts or aspects of a solution if functional requirements are coupled or become coupled in the design of products and processes.

(2) Integrate functional requirements into a single physical part or solution if they can be independently satisfied in the proposed solution.

(3) Minimize the number of functional requirements and constraints.

(4) Use standardized or interchangeable parts whenever possible.

(5) Make use of symmetry to reduce the information content.

(6) Conserve materials and energy.

(7) A part should be a continuum if energy conduction is important.

Many of these corollaries will be recognized as more concise statements of the foregoing DFM principles. In fact, all of the DFM principles and accompanying discussion given earlier can be derived or inferred from the design axioms. Hence, axiomatic design presents a useful heuristic approach to DFM. It states a few general rules that will always lead to good results and, as such, offers a way to proceed from the very general to the specific, rather than beginning with details. Most importantly, axiomatic design tends to improve the quality of early design decisions. By explicitly stating what is intuitively obvious to experienced designers and commonly used by good designers, the axioms provide a firm basis upon which design knowledge can be expanded in a systematic manner.

To illustrate how the design axioms provide insight, it is interesting to examine the distribution of information within a manufacturing system. In the past, all information was designed into hardware. In recent years, emphasis has been on transferring as much information as possible to software. As Yasuhara and Suh (1980) point out, the second axiom demands that the total information be minimized by judicious distribution between hardware and software. Hence, the best solution is often one that makes clever use of both mechanical devices and computers. In the case of maximizing assembly compliance discussed earlier, this insight is substantiated by the fact that compliance designed into both the product and tooling is preferable to a vision or tactile sensing solution. Similarly, the use of a simple SCARA arm and 'Z-stack' assembly is much preferable to robots having six degrees of freedom assembling products designed so that the components must be inserted sideways as well as vertically (Seering, 1985).

■ Quantitative evaluation methods

A second, very significant part of DFM has been the development of quantitative evaluation methodologies which allow the designer to rate the manufacturability of his design quantitatively. These methodologies provide systematic, step-by-step procedures, which ensure that the DFM rules are being correctly applied, encourage the designer to improve the manufacturability of his design, show the way by providing insight and stimulating creativity, and usually reward the designer with improved quantitative scores if he does well. An important added benefit of these methodologies is that they teach good design practice. Consequently, the need for repetitive use of the methodology diminishes with use.

At present, there are two quantitative evaluation methodologies in general use, both of which focus on ease of product assembly. Perhaps the best known and most widely used of these methods is the design for assembly method developed by G. Boothroyd and P. Dewhurst at the University of Massachusetts and, presently, at the University of Rhode Island. Step-by-step details of the methodology are presented in their *Design for Assembly – A Designer's Handbook* (1983a). Many technical papers by the handbook authors (too numerous in number to be included as references in this paper) have appeared in conference proceedings, trade journals, etc., describing the methodology and the basis behind it. The Boothroyd–Dewhurst DFA handbook contains three sections dealing with choice of assembly method, design for manual assembly and design for automatic assembly. An excellent overview of these sections is presented by the handbook authors in a series of articles published by *Machine Design* (Boothroyd and Dewhurst, 1983b; Boothroyd and Dewhurst, 1983c; Dewhurst and Boothroyd, 1984a). In a fourth *Machine Design* article (Dewhurst and Boothroyd, 1984b), the authors discuss design for robotic assembly. Corresponding software for personal computer systems has also been developed by the authors (Boothroyd and Dewhurst, 1983d; Dewhurst and Boothroyd, 1984b). The Boothroyd–Dewhurst DFA methodology has been adopted by a number of large US manufacturers including Ford Motor Co, Xerox Corp and Whirlpool Corp, to name a few.

The second quantitative methodology, known as the Hitachi assemblability evaluation method, AEM, is less widely used domestically because it is a proprietary method. Available from Hitachi Ltd, Tokyo, Japan, modified versions of this methodology are being used by General Electric Co. and General Motors Corp.

☐ Boothroyd–Dewhurst DFA method

Based largely on industrial engineering time study methods, the DFA method developed by Boothroyd and Dewhurst seeks to minimize cost of assembly within constraints imposed by other design requirements. This is done by first reducing the number of parts to be assembled and then ensuring that the remaining parts are easy to assemble. Essentially, the method is a systematic, step-by-step implementation of the DFM principles discussed earlier. To provide an overview of the method, the three sections in the DFA handbook (Boothroyd and Dewhurst, 1983a) are briefly described.

Choice of assembly method

The first section of the handbook provides a procedure for choosing between manual, special-purpose automatic, or programmable automatic assembly. Only basic information is needed for making a good estimate of the most economical assembly method. Knowledge of product design details is not necessary. Basic information required includes production volume per shift, number of parts in the assembly, single product or a variety of products, number of parts required for different styles of the product, number of major design changes expected during product life and the company investment policy regarding labour-saving machinery. The method for selecting the appropriate assembly method is contained on one colour-coded chart (Chart 1) given in the handbook. The chart is based on an analysis of mathematical models of the various assembly processes (Boothroyd and Dewhurst, 1983b). Using the basic information, the procedure is to choose an appropriate row and column. Intersection of the row and column defines a box on the chart containing a two-character abbreviation for the most economic assembly system. By varying the basic information slightly, different boxes are chosen thereby giving a feel for which parameters are driving the assembly method choice.

Design for manual assembly

The basic DFA evaluation procedure consists of comparing an 'ideal' assembly time with an estimated 'actual' assembly time required for a particular product design. To calculate the 'ideal' assembly time, the theoretical minimum number of parts is first determined by asking the following questions of each part in the assembly:

(1) Does the part move relative to all other parts already assembled?

(2) Must the part be of a different material than or isolated from all other parts already assembled?

(3) Must the part be separate from all other parts already assembled because otherwise necessary assembly and disassembly of other parts would be impossible?

If the answer to the part under consideration is 'yes', the number of parts is entered into the calculation; otherwise a 'zero' is assigned. The theoretical minimum number of parts is the sum of the numbers assigned to each part in the assembly. The 'ideal' assembly time is calculated assuming an assembly containing the theoretical minimum number of parts, each of which can be assembled in an 'ideal' time of 3 s. This ideal time assumes that each part is easy to handle and insert, and that about one-third of the parts are secured immediately upon insertion with well-designed snap-fit elements.

To estimate the 'actual' assembly time, penalties in seconds are assessed for handling difficulties and insertion difficulties associated with each actual part in the assembly. The penalties are based on a compilation of standard time study data as well as dedicated time study experiments. This data is tabulated as a function of part geometry, orientation features, handling features, method of attachment, etc., in the form of charts, one for manual handling (Chart 2) and one for manual insertion (Chart 3). 'Actual' assembly time is the sum of the handling and insertion times obtained from the charts for each part in the 'actual' assembly. The manual assembly design efficiency is computed as the ratio of 'ideal' assembly time to 'actual' assembly time.

Following evaluation, the assembly is redesigned for ease of assembly by first eliminating and combining parts using insights gained from the theoretical minimum number of parts determination. Following this, the remaining parts are redesigned to provide features which reduce assembly time, again using insights gained from the DFA analysis. To measure improvements in assemblability, the redesigned assembly can be analysed and the resulting efficiency compared with that of the old design. An important result of the DFA analysis is that it clearly shows that even products intended for manual assembly can benefit greatly if assemblability is considered early in the product design process.

Design for automatic assembly

The design for automatic assembly analysis consists of four steps:

(1) estimate cost of automated bulk handling and oriented delivery;

(2) estimate cost of automatic part insertion;

(3)   decide whether the part must be separate from all other parts in the assembly;

(4)   combine the results of steps 1–3 to estimate the total cost of assembly.

Although more computations are involved, basis for the design efficiency calculation and procedure for product redesign is essentially the same as for manual assembly. Cost penalties associated with ease of automatically feeding and orienting of individual parts is assessed based on consideration of part geometry (Charts 4, 5 and 6) and flexibility, weight, size, propensity to nest and tangle, etc. (Chart 7). Automatic workhead cost for part insertion is estimated based on classification of the insertion processes involved (Chart 8).

■ Computer-aided DFM

One approach to implementing DFM is to use an appropriate set of principles and rules to help guide the design of the product and then to evaluate and redesign the product using an appropriate evaluation methodology. To assist this process, a third DFM thrust has been the development of a variety of computer-based and/or computer-aided design programs. These programs may or may not interface with a CAD system and many are proprietary software products. Developments in this area include commercially available CAD software, research involving conventional interactive computer programming approaches and research involving AI/expert system approaches.

☐   Commercial software

A variety of commercial software has become available which provides DFM assistance to the designer. Programs which assist in the design of individual parts for a particular fabrication process are most common. One example of this type of program is Moldflow, a product of Moldflow Australia (PTY) Ltd. Moldflow is a computer simulation of molten plastic moving through the gates, runners and cavity of an injection mould (Kuttner, 1985). Licensed to Ford for use at a specified number of user sites, Moldflow has been interfaced to Ford's product design graphics system's finite element module. Embodied within the program is a DFM philosophy that encourages mouldability analysis by pointing to part features that might cause warping and failure in production. The program does this, not by telling the designer how to design the part or where to gate the mould, but by indicating results to be expected from a given choice of design and processing parameters. By performing 'what–if' variations of his design, the designer is able to converge iteratively to the best solution.

Variation simulation analysis software, better known as VSAS, is an example of another type of commercial software which embodies and facilitates the DFM philosophy. The VSAS package, marketed by Applied Computer Solutions in St Clair Shores, MI, allows the designer to predict assembly tolerance and manufacturing variation before prototype build. This is done by creating a model which includes the size and variation of each component and accuracy of each assembly operation based on tolerances specified by the designer. Using a Monte Carlo simulation procedure, VSAS simulates a production run by putting the assembly together, one step at a time, in the proper processing sequence, a specified number of times. Results of the simulation are analysed, and a complete statistical picture of the proposal process is provided including a population distribution of critical dimensions, high and low limits, percentage of out-of-spec parts, and percent contribution of each component and operation to final assembly tolerances. Like Moldflow, the VSAS software facilitates DFM by enabling the designer to minimize overall dimensional variation through experimentation with tolerances and datums using 'what–if' optimization. Because the model requires input and interaction from both design and processing, use of the program also fosters team building, an essential part of the DFM philosophy. The VSAS software package runs on the IBM PC as well as a variety of mainframes. It also accepts part geometry data directly from most popular CAD systems.

A major barrier to DFM is usually time. Product designers are typically operating under very tight schedules and are therefore reluctant to spend time considering DFM issues. Computer software which simplifies the effort and shortens the time required to implement DFM on a daily basis represents a third type of commercial DFM software product. A good example of this is the Boothroyd–Dewhurst DFA software package (Boothroyd and Dewhurst, 1983d; Dewhurst and Boothroyd, 1984b), available from the authors. This package is an interactive, computerized version of the Boothroyd–Dewhurst DFA method discussed above. It consists of six program modules available for IBM PC or compatible computers. The modules include assembly system economics, assembly machine simulation, design for manual assembly, design for automatic assembly, design for automated handling and design for robotic assembly (new in 1985). Use of the software simplifies data entry, manipulation and reporting, freeing the designer to concentrate on the DFA evaluation and the innovation it stimulates.

☐ Research developments

Dargie, Parmeshwar and Wilson present an interesting approach to computer-aided DFM called MAPS-1 (1980). Recognizing that material and process alternatives should be carefully considered early in the design process, well before part geometries are specified, MAPS-1 provides a short list of the best combinations for further consideration by the designer. In the words of the authors: 'The MAPS-1 system is intended as a general purpose aid to the designer in making preliminary selections of materials and manufacturing processes for a given part.' The system uses a 12-digit classification code to interactively capture essential features of the part under consideration. The code is then used in conjunction with material and processing databases progressively to eliminate materials and processes, beginning with obviously unsuitable choices, and then proceeding to incompatible or difficult material and process combinations. The material/process combinations which remain are divided into two categories, usual practice and unusual practice. If the list of material/process combinations which remain is too large to be easily evaluated, the user may elect to have the program rank each candidate according to the predetermined criterion. Several difficulties are inherent in the MAPS-1 approach. Significant among these is the need to differentiate between primary and secondary processing; difficulties in dealing with process chains and processes such as heat treatment and surface coating which do not contribute to part geometry; developing and properly representing the large amounts of data required for the databases; and manipulating and searching the databases involved.

Use of the concept of 'optimal suggestions' developed by Jakiela, Papalambros and Ulsoy (1985) is another interesting approach to computer-aided DFM. In this approach, the authors accept that creative synthesis, or the design concept phase, will remain a human task for some time to come, and therefore ask what can be done to enhance the designer's capabilities in that stage. Their solution consists of creating a program which makes 'suggestions' to the designer during the conceptualization of the design. The suggestions are formulated in such a way that if they are all followed, an optimal solution will be achieved. Hence, the suggestions act both to stimulate creativity and to show the way to good design. To illustrate this approach, optimal suggestions based on the Boothroyd–Dewhurst design for automatic assembly method (1983a) are programmed. The suggestions ask the designer to design the individual parts of the product in such a way that a high assemblability efficiency will be attained. If other design constraints make it impossible to follow a suggestion, then the next best design is pursued. A major research challenge posed by this approach involves programming needs which are generally different from those of usual programming for numerical data manipulation.

Wood, Cohen, Medeiros and Goodrich have developed a computerized approach to design for robotic assembly (1986), which aids in the design process either from the initial conceptual stage or for redesign. The methodology described seeks to minimize the number of parts used to achieve all required product functions while producing a design requiring minimum assembly cost in the form of robot and special tooling. This is done by guiding the user through the product design process in such a way that he must consider and deal with specific robotic assembly issues as he develops the product design concept. Guidance is based on numerous design principles which facilitate robotic assembly. The methodology is implemented as an interactive package in Microsoft FORTRAN for the IBM Personal Computer and consists of five sections. In the first section, the major components needed to satisfy the product's functional requirements are defined. Fasteners and other secondary parts are defined in a later section. The second section guides the designer towards solutions which avoid floppy components. Identification of self-fixturing components and part assembly sequence is accomplished in Section 3. Section 4 deals with definition of part interface and fastening methods based on consideration of gripper access, positioning accuracy, orientation and placement stability, and robotic capabilities required. The fifth and final section allows the user to examine output information generated by the program in a variety of ways. By studying the output information, the user can better understand the design problem at hand and iterate quickly to a desirable solution.

☐ AI/expert systems

As illustrated by some of the computer-aided DFM developments discussed thus far, the large amount of principles, rules, guidelines and other heuristic data inherent in the DFM approach leads to a variety of difficulties when conventional computer programming techniques are employed. The field of artificial intelligence, AI, and expert systems embodies a range of new programming techniques which appear to be well suited to DFM programming needs. As indicated in a review of expert systems applications in mechanical design given by Dixon and Simmons (1983), however, these possibilities are just beginning to be explored. Findings of a workshop addressing the use of AI in design (Cole *et al.*, 1986) indicate that extensive work on knowledge representation as well as development of AI techniques which avoid shortcomings associated with rule-based expert systems currently being used with success in other applications is needed to facilitate meaningful application of AI to design. Research programs at the University of Massachusetts under the direction of John Dixon and at the Carnegie–Mellon Intelligent Systems Laboratory directed by Mark Fox are of particular interest because they focus specifically on DFM issues.

A recent discussion of the expert system for mechanical design research program at the University of Massachusetts is given by Dixon and Simmons (1985). Top-level goal of this program is to develop a theory and practice for mechanical design and manufacturing processes. In the words of the authors, specific objectives include:

(1)   to learn how to develop expert systems in CAD environments that can do on-line evaluation of designs for their manufacturability;

(2)   to explore the use of design with features as a means for creating a design database that will serve manufacturing process planning as well as design and analysis needs;

(3)   to develop a new language for knowledge representation in design that will facilitate the construction of expert systems in mechanical design.

To implement this research program, a series of subproject topics have been selected for use in gaining the theoretical understanding and practical experience needed to achieve the desired research objectives. Some of the specific topics under consideration include design of heat sinks, design and analysis of injection-moulded parts, design of plastic extrusions, casting design, analysis, and process selection, plastic materials selection, and a domain-independent iterative redesign program. In outlining their program of research, the authors note that an important research result is likely to be the creation of single databases useful for both design and manufacturing.

The Intelligent Systems Laboratory at Carnegie–Mellon University, CMU, under the direction of Mark Fox, is presently initiating research in constraint-directed reasoning applied to the design process. Called 'design fusion', the ideas involved in this project were discussed recently by M. Rychener of CMU (Cole *et al.*, 1986). The goal of design fusion is to eliminate engineering change, that is, to do the design right the first time. The constraint-directed reasoning approach asserts that all needs and requirements which occur 'downstream' from design (that is, manufacturing and assembly, testing, distribution, service, etc.) can be represented as constraints on the design. The problem of research then is to discover how best to formulate these constraints, many of which are coupled in complex ways, so that they properly guide the designer. If successful, the design fusion research project will generate the theories and techniques required by a computer-aided design system to assist the design process in optimizing performance of downstream activities.

■ Concluding remarks

Design for manufacture is a large subject spanning many disciplines and points of view. Complete comprehension of the many complex and interrelated issues encompassed by DFM is still far in the future. Yet an urgent need for proper application of this understanding exists today. DFM is recognized as the key to simultaneously minimizing manufacturing cost, assuring product quality and realizing the productivity increase promised by advanced manufacturing technology. Ultimately, DFM is a philosophy which forms the basis of a common language between research, design and manufacturing as well as all other facets of the manufacturing system.

References

Altamuro V. (1985). *Design for Robotic and Automatic Assembly*. Seminar notes, ASME Des. conf. and show, by Robotics Research Consultants, PO Box 3276, Toms River, NJ

Andreasen M. M., Kahler S. and Lund T. (1983). *Design for Assembly*. Bedford, UK: IFS Publications

ASCO Press (1985). *Powder Metallurgy (P/M) Newsletter*. ASCO Sintering Co, 2750 Garfield Ave, Commerce, CA 90040, Sept

Boothroyd G. and Dewhurst P. (1983a). *Design for Assembly – A Designers Handbook*. Dep. of Mech. Eng. Univ. of Massachusetts, Amherst.

Boothroyd G. and Dewhurst P. (1983b). Design for assembly: Selecting the right method. *Machine Des.* (10 Nov), 94–8

Boothroyd G. and Dewhurst P. (1983c). Design for assembly: Manual assembly *Machine Des.* (8 Dec), 140–5

Boothroyd G. and Dewhurst P. (1983d). Computer-aided design for assembly. *Assembly Eng.* (Feb)

Bradyhouse R. G. (1984). Design for assembly and value engineering helping you design your product for easy assembly. In *SAVE Conf. Proceedings*, Society of American Valve Engineers, Irving, TX, 14–23

Chow W. W. C. (1978). *Cost Reduction in Product Design*. New York: Van Nostrand Reinhold

Church F. I., ed. (1981). Impacts: Light, tough, precise; reduce machining; replace assemblies. *Mod. Metals* (Mar), 18–24

Cole J. H., Stoll H. W. and Parunak H. V. D. (1986). Machine intelligence in machine design: Workshop summary. *Artificial Intelligence Eng.*, **1**(1), 54–7

Dargie P. P., Parmeshwar K. and Wilson W. R. D. (1980). *MAPS-1: Computer-aided Design System for Preliminary Material and Manufacturing Process Selection*. ASME paper 80-DET-51, presented at the ASME Des. Eng. Tech. Conf., Beverly Hills, CA, Sep–Oct

Dewhurst P. and Boothroyd G. (1984a). Design for assembly: Automatic assembly. *Machine Des.* (26 Dec), 87–92

Dewhurst P. and Boothroyd G. (1984b). Design for assembly: Robots. *Machine Des.*, **58**(4) (23 Feb), 72–6

Dixon J. R. and Simmons M. K. (1983). Computers that design: Expert systems for mechanical engineers. In *Computers in Mechanical Engineering*. ASME, New York, Nov, 10–18

Dixon J. R. and Simmons M. K. (1985). *Expert Systems for Mechanical Design: A Program of Research*. ASME paper 85-DET-78, Sept

Ettlie J. E. and Reifus S. A. (1986). *The Integration of Design and Manufacturing for Deployment of Advanced Manufacturing Technology*. Presented at the TIMS/ORSA joint national meeting, April, Bonaventure Hotel, Los Angeles, CA

Gage W. L. (1967). *Value Analysis*. London: McGraw-Hill

General Electric (1970). *Manufacturing Producibility Handbook*, Schenectady, NY: General Electric

Greenwood D. C. (1959). *Product Engineering Design Manual*. New York: McGraw-Hill

Howell V. W. (1982) *Are Producibility and Productivity Correlated?* SME Tech paper

AD82-153, presented at the Assembly VIII Conference, Cleveland, OH. 16–28 Mar, Society of Manufacturing Engineers, Dearborn, MI

Huthwaite B. (1985). *Product Design for Automated Assembly*. Workshop notes, Michigan State Univ., Management Education Center, Troy, MI

Jakiela M., Papalambros P. and Ulsoy A. G. (1985). Programming optimal suggestions in the design concept phase: Application to the Boothroyd assembly charts. *J. Mech. Transmiss. Autom. Des.*, **107**(2), 285–91

Knill B., ed. (1985). Allen–Bradley puts its automation where its market is. *Mat. Handling Eng.*, **40**(7) (Jul), 62–6

Koenig D. T. (1981). *Interrelationships between Methods Engineering and Producibility Engineering*. ASME paper 81-DE-3, presented at the Des Eng Conference and Show, Chicago, IL, 27–30 Apr

Kuttner B. C. (1985). Computer-aided plastics engineering. *Manufact. Eng.*, **92** (Mar), 90–1

Laszcz J. F. (1984). Product design for robotic and automatic assembly. In *SME Robots 8 conference proceedings. Applications For Today*, Jun, Vol 1

Lewis G. (1986) Designing to reduce assembly costs. *Plas Des. Forum*, **11**

Lotter B. (1984) Using the ABC analysis in design for assembly. *Assembly Autom*, **4**(2), 80–6

Mudge A. E. (1971). *Value Engineering, A Systematic Approach*. New York: McGraw-Hill

Nevins J. L. and Whitney D. W. (1982). *Programmable Assembly System Research and its Application – A Status Report*. SME Tech Paper MS82-125, Society of Manufacturing Engineers, Dearborn, MI

Pierson R. A. (1985). IBM's automated material handling system. *Robotics World*, **3**(2) (Feb), 20–3

Precision Metal (1975). Design for assembly. Jul, 26–8

Riley F. J. (1983). *Assembly Automation: A Management Handbook*. New York: Industrial Press

Schreiber R. R., ed. (1985). Design for assembly. *Robotics Today*, **7**(3). 45–6

Seering W. P. (1985). Who said robots should work like people? *Tech. Rev.*, Apr, 59–67

Stoll H. W. (1986). *Product design for efficient manufacture*. Workshop notes, Indust. Tech. Inst., Ann Arbor, MI

Suh N. P., Bell A. C. and Gossard D. C. (1978). On an axiomatic approach to manufacturing and manufacturing systems. *J. Eng. Indust.*, **100**(2), 127–30

Tuffentsammer K. (1979). Economical design and design for economic manufacture. In *Research-developing-Design*. RKW-Handbook, Erich Schmidt-Verlag, Berlin

Wood B. O., Cohen P. H., Medeiros D. J. and Goodrich J. L. (1986). Design for robotic assembly. *Robotics and Manufacturing Automation*, PED, Vol. 15, ASME book No. G00321

Yasuhara M. and Suh N. P. (1980). A quantitative analysis of design based on axiomatic approach. In *Computer Applications in Manufacturing Systems*. ASME Prod. Eng. Div. Publ., PED, Vol. 2

3.2
A methodology for planning experiments in robust product and process design

Anne C. Shoemaker and Raghu N. Kacker

AT&T Bell Laboratories, Holmdel, NJ 07733, USA

■ Introduction

In 1980 Genichi Taguchi introduced his approach to using statistically designed experiments in robust product and process design to US industry. Since then, the robust design problem and Taguchi's approach to solving it has received much attention from product designers, manufacturers, statisticians and quality professionals, and many companies have applied the approach (Phadke *et al.*, 1983; Pao *et al.*, 1985; Lin and Kacker, 1985; Byrne and Taguchi, 1986; Kacker and Shoemaker, 1986; Phadke, 1986; American Supplier Institute, 1984, 1985 and 1986). Most agree that Taguchi poses a very important problem: the design of products and manufacturing processes which perform consistently on target and are insensitive to hard-to-control variations.

The particular methods Taguchi employs to solve this problem are more controversial, however. This controversy has led to a proliferation of research to better understand the problems in robust product and process design, and to provide improved methods to solve these problems.

Although the answers are not all in yet, the importance of the problem

This paper is reproduced by permission of John Wiley and Sons Limited from *Quality and Reliability Engineering International*, Vol. 4, No. 2 (1988), pp. 95–103. © 1988 by John Wiley and Sons Limited has led some statisticians and engineers at AT&T Bell Labs to develop a methodology implementing what we feel are justifiable and practical techniques based on our current understanding. We expect the techniques to evolve as our understanding of the problems and their appropriate solutions develops. However, the current methodology and the way it is presented has been successful in significantly increasing engineers' independent and successful use of robust product and process design methods and statistically designed experiments. This paper concentrates on the problem formulation and experiment planning stages of the methodology.

An important feature of the robust design methodology is its operational nature. The methodology organizes the decisions, trade-offs, and judgements that are so important in experiment planning and analysis into an explicit step-by-step strategy. At each step, tools such as flowcharts, tables and graphs are provided to simplify the experiment planning process and help engineers make important decisions.

The approach we are taking, especially in experiment planning, is bold. For example, we encourage study of many parameters in a small, highly fractionated experiment. When a parameter is expected to have a non-linear (especially non-monotone) effect on the response or when two parameters are suspected to interact, the experiment should allow for those non-linearities. If possible, the experiment should also allow for unexpected non-linearities, but usually not if this would require more experimental runs or dropping parameters. Under this strategy, the confirmation experiment plays a crucial role, checking that assumptions made in planning the main experiment were not badly violated.

This robust design methodology is taught to product designers, manufacturing process designers and manufacturing engineers in a three-day workshop. In this workshop, about half of the time is spent in lectures on the basic objectives, concepts and methods of robust design. The other half is spent partly on exercises, but mostly in applying material learned in the morning to real projects from the students' jobs. At the end of the workshop, students have defined their problem and objectives, planned an experiment and documented their objective and plan. On the last day, they orally present their objectives and plans to their management.

The step-by-step methodology and the format used to teach it have been effective in attaining a moderate level of robust design and design of experiments application at AT&T. There were approximately 75 applications of this methodology in 1986. The application rate has been almost doubling every year since 1980.

We begin by describing the robust design problem. In the next section we give a high-level view of the complete methodology, and describe how the methodology is taught to engineers in the three-day workshop. In the final two sections we focus in detail on the methodology's problem formulation and experiment planning steps. In describing these steps we also compare the approach we are recommending with alternative approaches.

■ The basic robust design problem

Robust design is the operation of choosing settings for product or process parameters to reduce variation of that product or process's response from target. Because it involves determination of parameter settings, robust design is called 'parameter design' by Taguchi and others (Kacker, 1985; Taguchi, 1986; Phadke, 1982).

Figure 1 shows a block diagram representation of a simple robust design problem. The block represents the product or process under study. That product or process's responses are determined by a large number of variables. Some of these variables are under the designer's control, and these are called *control parameters*. Examples of control parameters include temperature or flow-rate settings in a manufacturing process, and nominal resistances of components in an electronic circuit. An assumption usually made in robust design is that the cost of producing the product at each level of any given control parameter is the same.

The responses are also influenced by variables which are difficult or expensive for the designer to control. These are called *noise variables*. Examples of noise variables include typical manufacturing variations such as non-uniformity in raw materials, deviation of components from their nominal specifications, as well as variation in the customer's environment, and deterioration or wear-out in component parts over time.

In theory, some of these noise variables could be controlled. For example, the designer could control raw material variations by specifying a higher-grade raw material or more expensive components with higher tolerances. Taguchi calls this activity *tolerance design* (Kacker, 1985; Taguchi, 1986). However, these efforts to reduce response variation by controlling noise variables increase the cost of each unit of product produced.

In contrast, robust design is a way to reduce response variation by using the controllable parameters to dampen the effects of the hard-to-control noise variables. This approach does not increase the cost of the product. If application of robust design does not sufficiently reduce response variation, designers can further reduce this variation using tolerance design.

For more discussion of the objectives and benefits of robust product and process design, see Kacker (1985 and 1986).

In some problems, such as circuit design or thermal design, a computer model is available which relates the control parameters and noise variables to the response characteristic. In these cases, the objective of robust design can be accomplished using non-linear optimization techniques and the computer model.

This paper addresses the situation where knowledge of the functional relationship between response, control parameters and noise variables is incomplete. In this situation, physical experiments are needed to investigate this relationship and identify improved control parameter settings.

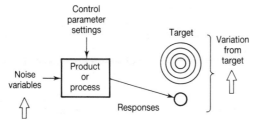

- Manufacturing variation
- Component tolerances
- Customer use conditions
- Deterioration

Figure 1   Block diagram representation of a simple robust design problem.

■ A four-step robust design methodology

In AT&T, as in many companies, on-going job-related engineering education is primarily accomplished through short courses lasting from two to five days. In this limited time, the challenge is to provide engineers with a problem-solving methodology that they can successfully apply back on their jobs.

In these situations, flowcharts, steps and road maps are valuable ways to give structure to the tasks and decisions that are part of a problem-solving methodology. Figure 2 shows the highest level flowchart for the robust design methodology. In the 'formulate the problem' step, an engineer prepares for experiment planning by clearly defining quality improvement goals and listing and classifying variables that may determine achievement of those goals. As we will see in the next section, the experimenter makes a prioritized list of parameter effects to be estimated from the experiment results. The experiment plan is constructed with the aid of special tools which we describe in the final section.

Once the experiment is conducted, the results are analysed to identify key variables and the best settings for these variables. The main tools for this step are simple plots of average response and variability measures for each setting of each parameter, and half-normal probability plots to identify key parameters.

Because the experiment planned in step 2 is usually highly fractionated, the fourth step, confirmation of improvement, is very important. This small follow-up experiment consists of at least two test combinations: one at the new parameter settings identified in step 3, and one at the original parameter settings.

Kacker and Shoemaker (1986) show how these steps were followed to improve the design of a process for growing epitaxial layers on silicon

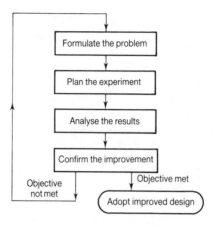

Figure 2   The four major steps in the robust design methodology.

wafers. In the remainder of this paper we focus in detail on the problem formulation and experiment planning steps.

■ Step 1: formulate the problem

In this step engineers define the robust design problem by clearly stating their objectives for the product or process improvement, and specify product/process response characteristics that reflect these objectives. They also list control parameters and noise variables, and make tentative decisions about non-linearity that should be allowed for in the experiment.

The 'formulate the problem' step is often completed by a group of engineers in a brainstorming session. If this group includes all the important players the experiment will have a much better chance of being completed and successful. For process design problems, the key players typically include process designers and manufacturing engineers responsible for set-up and maintenance of the production line, and product designers if changes to the product design might help solve processing problems.

It is very important to measure continuous, 'fundamental' characteristics, as opposed to discrete characteristics such as yield. Phadke and Taguchi (1987) illustrate the importance of choosing a good response characteristic in several examples.

Unfortunately, especially in manufacturing organizations, insufficient resources are often devoted to obtaining good in-process measurements. Instead, measurements are taken too far downstream in the process, so that the results of a single process step are difficult to isolate from the effects of later process steps. Improvement of measurements should be a top priority for manufacturers who aim to produce high-quality, low-cost products.

After deciding upon the response characteristics to be measured, engineers list the product or process variables which affect the characteristics and classify these variables as controllable ('control parameters') or difficult-to-control ('noise variables'). In manufacturing process design, controllable parameters are typically the *nominal* values of process parameters, and noise variables often include variation or drift of the *actual* values of these process parameters from their nominal settings, variation in raw materials and component parts, and variation in the results of previous process steps.

In the 'formulate the problem' step, engineers also list the pairs of control parameters which they expect to interact in their effect on one or more response characteristics.

After completing these lists, engineers decide on a tentative number of settings (usually two, three or four) for each control parameter. When the parameter is categorical, this decision is usually not difficult. When the parameter is continuous, testing at three or four settings, rather than just two settings, is considered if any of the conditions illustrated in Figure 3 hold. Figure 3(a) illustrates a scenario which may be quite common in robust design applications. Often, a multi-parameter experiment is resorted to after one or more one-parameter studies have failed to achieve the desired level of performance. It is quite possible that a parameter which has already been 'tuned' is now near its optimum. Testing a previously tuned parameter at three settings tries to ensure against this scenario.

Figures 3(b) and 3(c) show situations in which a great range of settings is possible for a parameter, and when the direction of improvement is not definitely known. In these cases, three settings help ensure that non-monotone behaviour will be detected. Figure 3(d) illustrates a situation in which the direction of improvement is known, but it is not known how far out the improvement will continue before it gradually or suddenly degrades.

To facilitate trade-off decisions that may be necessary in planning the experiment, the lists of control parameters, interactions between control parameters and noise variables are prioritized.

■ Step 2: plan the experiment

The plan for a robust design experiment has two parts, a 'control array' and a 'noise plan', as shown for example in Figure 4. As its name indicates, the control array is the plan for varying the settings of the control parameters in the experiment. The rows of this array correspond to different configurations of the product or manufacturing process. The noise plan is the scheme for measuring the effects of noise variables. This plan might be an array, called a

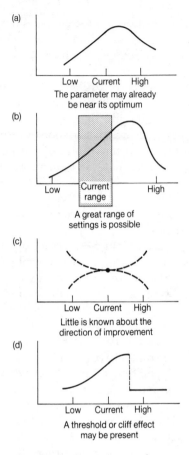

(a)

Low   Current   High

The parameter may already
be near its optimum

(b)

Low   Current   High
             range

A great range of
settings is possible

(c)

Low   Current   High

Little is known about the
direction of improvement

(d)

Low   Current   High

A threshold or cliff effect
may be present

Figure 3   Conditions for testing at three or more levels.

'noise array', specifying how noise variables will be explicitly varied in the experiment. In physical experiments in manufacturing process design, however, the noise plan is often a plan for taking multiple measurements of the response characteristic in a manner designed to capture the effect of underlying noise variables.

In order that fair comparisons can be made among the rows of the control array the same noise plan is repeated for each row.

☐   **Step 2A: construct the control array**

Orthogonal arrays

To facilitate construction of the control array, we use several tools. The first tool is the orthogonal array, OA, a table of integers whose columns are pairwise balanced. That is, in any two columns every ordered pair of integers

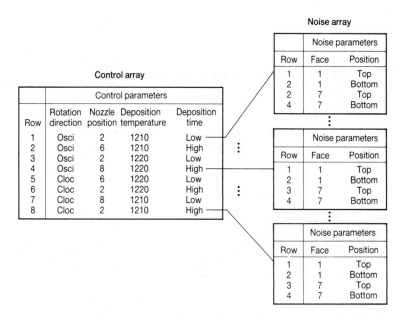

Noise array

| | | Noise parameters | |
Row	Face	Position
1	1	Top
2	1	Bottom
2	7	Top
4	7	Bottom

Control array

| | | Control parameters | | |
Row	Rotation direction	Nozzle position	Deposition temperature	Deposition time
1	Osci	2	1210	Low
2	Osci	6	1210	High
3	Osci	2	1220	Low
4	Osci	8	1220	High
5	Cloc	6	1220	Low
6	Cloc	2	1220	High
7	Cloc	8	1210	Low
8	Cloc	2	1210	High

| | | Noise parameters | |
Row	Face	Position
1	1	Top
2	1	Bottom
3	7	Top
4	7	Bottom

| | | Noise parameters | |
Row	Face	Position
1	1	Top
2	1	Bottom
3	7	Top
4	7	Bottom

Figure 4 A typical robust design experiment plan. The noise array is repeated for each row in the control array.

occurs the same number of times. Five orthogonal arrays are shown in Figure 5.

Although much of the discussion of Taguchi's methods has focused on his use of orthogonal arrays, these arrays were not invented by Taguchi. All regular m^{k-p} fractional factorial experiment plans are orthogonal arrays. For example, OA_8 is a 2^{7-4} fractional factorial plan, and OA_{16} is a 2^{15-11} fractional factorial plan. OA_{12} is a Plackett–Burman plan. Rao (1947) gives a general theory of orthogonal arrays (which he calls hyper-Graeco Latin squares), and others have studied their construction and catalogued useful sets of orthogonal arrays (see, for example, Raghavarao (1971) and Kacker (1988)). Although there are many orthogonal arrays, only seven or eight of them are commonly used to construct plans for physical experiments (Taguchi, 1986).

To construct the control array, control parameters are assigned to the columns of an orthogonal array, and the integers in these columns are translated into the actual test settings of the assigned parameters. The unassigned columns are deleted from the array.

In assigning control parameters to columns of some orthogonal arrays the experimenter can allow for certain interactions by choosing the columns for assignment carefully. For example, in OA_8, column 3 corresponds to the interactions between columns 1 and 2 and between columns 4 and 7. So if parameters A, B, C and D are assigned to columns 1, 2, 4 and 7, respectively, a fifth parameter should not be assigned to column 3 if A and B or C and D are likely to interact.

OA4 Column		
Row	1 2 3	
1	1 1 1	
2	1 2 2	
3	2 1 2	
4	2 2 1	

OA8 Column	
Row	1 2 3 4 5 6 7
1	1 1 1 1 1 1 1
2	1 1 1 2 2 2 2
3	1 2 2 1 1 2 2
4	1 2 2 2 2 1 1
5	2 1 2 1 2 1 2
6	2 1 2 2 1 2 1
7	2 2 1 1 2 2 1
8	2 2 1 2 1 1 2

OA9 Column	
Row	1 2 3 4
1	1 1 1 1
2	1 2 2 2
3	1 3 3 3
4	2 1 2 3
5	2 2 3 1
6	2 3 1 2
7	3 1 3 2
8	3 2 1 3
9	3 3 2 1

OA16 Column	
Row	1 2 3 4 5 6 7 8 9 10 11 12 13 14 15
1	1 1 1 1 1 1 1 1 1 1 1 1 1 1 1
2	1 1 1 1 1 1 1 2 2 2 2 2 2 2 2
3	1 1 1 2 2 2 2 1 1 1 1 2 2 2 2
4	1 1 1 2 2 2 2 2 2 2 2 1 1 1 1
5	1 2 2 1 1 2 2 1 1 2 2 1 1 2 2
6	1 2 2 1 1 2 2 2 2 1 1 2 2 1 1
7	1 2 2 2 2 1 1 1 1 2 2 2 2 1 1
8	1 2 2 2 2 1 1 2 2 1 1 1 1 2 2
9	2 1 2 1 2 1 2 1 2 1 2 1 2 1 2
10	2 1 2 1 2 1 2 2 1 2 1 2 1 2 1
11	2 1 2 2 1 2 1 1 2 1 2 2 1 2 1
12	2 1 2 2 1 2 1 2 1 2 1 1 2 1 2
13	2 2 1 1 2 2 1 1 2 2 1 1 2 2 1
14	2 2 1 1 2 2 1 2 1 1 2 2 1 1 2
15	2 2 1 2 1 1 2 1 2 2 1 2 1 1 2
16	2 2 1 2 1 1 2 2 1 1 2 1 2 2 1

OA18 Column	
Row	1 2 3 4 5 6 7 8
1	1 1 1 1 1 1 1 1
2	1 1 2 2 2 2 2 2
3	1 1 3 3 3 3 3 3
4	1 2 1 1 2 2 3 3
5	1 2 2 2 3 3 1 1
6	1 2 3 3 1 1 2 2
7	1 3 1 2 1 3 2 3
8	1 3 2 3 2 1 3 1
9	1 3 3 1 3 2 1 2
10	2 1 1 3 3 2 2 1
11	2 1 2 1 1 3 3 2
12	2 1 3 2 2 1 1 3
13	2 2 1 2 3 1 3 2
14	2 2 2 3 1 2 1 3
15	2 2 3 1 2 3 2 1
16	2 3 1 3 2 3 1 2
17	2 3 2 1 3 1 2 3
18	2 3 3 2 1 2 3 1

Figure 5   Five commonly used orthogonal arrays.

In general, the only orthogonal arrays which can be used to allow for interactions are those corresponding to regular m^{k-p} fractions of factorial designs. These are OA_4, OA_8, OA_9, OA_{16}, OA_{27}, OA_{32}, etc. As we will see later, some other orthogonal arrays can allow for a very limited set of interactions.

Interaction graphs

When the experimenters wish to allow for interactions, the interaction graph can be a useful tool for simplifying the column assignment process (Kacker and Tsui, 1988). The interaction graph for OA_8 is shown in Figure 6. The numbers on the graph correspond to columns in the OA. A line segment connecting two nodes corresponds to the interaction between those nodes.

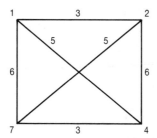

Figure 6 The interaction graph for OA$_8$.

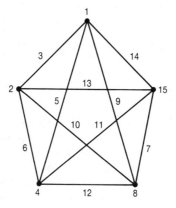

Figure 7 A linear graph for OA$_{16}$.

Hence, column 6 corresponds to the interactions between columns 1 and 7 and between columns 2 and 4, etc. Interaction graphs also exist for OA$_{16}$ and all other 2^{k-p} orthogonal arrays (that is, OA$_{32}$, OA$_{64}$, etc.).

To make column assignments using an interaction graph, assign parameters to nodes first, then when the nodes are used up, assign the remaining parameters to line segments. As you make assignments, cross out the line segments corresponding to interactions that are likely to be important. Notice from Figure 6 that when parameters are assigned only to nodes and not to line segments, no parameter main effects will be confounded with interactions between pairs of parameters. However, these pairwise interactions will be confounded with each other since interaction columns, such as column 3, appear more than once in the graph. This is a Resolution IV experiment plan (see Box *et al.* (1978), page 385 for a definition of resolution).

In addition to interaction graphs, there are other graphs that are sometimes useful for making parameter assignments. Taguchi (1986) gives a collection of 'linear graphs' for each commonly used orthogonal array. For example, Figure 7 shows a linear graph corresponding to OA$_{16}$. This graph is useful when the experimenter can afford 16 runs and there are five or fewer parameters. Notice that when parameters are assigned only to nodes in this graph the experiment plan will not confound pairwise interactions with each other. This is a Resolution V plan.

Tailoring an orthogonal array

It is often not possible to match the control parameter list made in the 'formulate the problem' step exactly to an orthogonal array. For example, the problem might require that one control parameter be tested at three settings, and four control parameters be tested at two settings. None of the standard orthogonal arrays in Figure 5 or given by Taguchi (1986) has one or more three-level columns and four two-level columns. Non-standard situations like this one can be accommodated by tailoring an orthogonal array to suit the problem at hand, using the 'combining columns' and 'dummy level' techniques.

The combining columns technique allows a four-setting parameter to be assigned to a two-setting array such as OA_8 or OA_{16}. Simply treat the four-setting parameter as if it were two two-setting parameters that interact, and use an interaction graph or linear graph to make the column assignments as above. For example, a four-setting parameter, A, can be accommodated in OA_8 by assigning it to columns 1 and 2. The levels of A are then obtained from the ordered pairs of integers in columns 1 and 2, as follows: $(1,1)$ becomes level 1 of A, $(1,2)$ becomes level 2 of A, $(2,1)$ becomes level 3 of A and $(2,2)$ becomes level 4 of A. Column 3 must be set aside because it corresponds to the interaction between columns 1 and 2, and is therefore now confounded with the main effect of A.

The collapsing level technique converts a k-level column into a $(k-1)$-level column by simply replacing one of the integers in the column by a single one of the others. For example, a four-level column becomes a three-level column by replacing every '4' by a '2'. The resulting column has twice as many 2s as 1s and 3s. The collapsing level technique is described by Addelman (1962).

These tailoring techniques ensure that parameter main effects are unconfounded and uncorrelated with each other because the control array's columns remain proportionally balanced (Addelman, 1962).

In addition to these techniques, Taguchi uses several more which are designed to squeeze more parameters into the array. The 'combination factor' method assigns two two-level parameters to a single three-level column in an orthogonal array such as OA_{18} (see Phadke *et al.* (1983) for an example of this kind of assignment). The 'idle column' method allows two four-level columns to be created by combining two pairs of two-level columns which share the same interaction column. For example, two four-level parameters, A and B, can be accommodated in OA_8 by combining columns 1 and 2 and also combining columns 4 and 7. However, since these pairs of columns both have column 3 as an interaction column, the main effects of parameters A and B will be confounded.

We do not include the idle column or the combination factor techniques in the basic methodology because they may correlate and confound parameter main effects.

To make the column assignment and tailoring process easy, it is helpful to assign the control parameters to the interaction graph or linear graph in a certain order. As a general rule, we recommend the following order for making column assignments:

(1) Assign parameters with more than two settings.

(2) Assign parameters which may interact, crossing out line segments corresponding to those interactions.

(3) Assign the remaining parameters.

If all the nodes are used up in stage 3, assign parameters to line segments. If all the nodes and line segments are used up, reduce the number of settings for some parameters, drop interactions, or, as a last resort, consider using a larger OA or dropping some control parameters from the experiment.

Comparison with alternative approaches for constructing experiment plans

It is interesting to compare the orthogonal array approach for constructing an experiment plan to the approach taken by Box *et al.* (1978, p. 410). Since a full orthogonal array represents a completely saturated experiment, when you take the OA approach you are starting with a saturated experiment and reducing the saturation by deleting columns from the orthogonal array. In contrast, Box *et al.* start by constructing a full factorial plan for a subset of the parameters, and then add more parameters to the plan by adding columns. Thus they are incrementally increasing the plan's saturation as they add columns. The orthogonal array approach to planning an experiment and the Box *et al.* approach could lead to exactly the same experiment plan.

If all control parameters are continuous-valued, central composite designs (see Box *et al.*, Chapter 15) offer an alternative strategy which provides estimates of a quadratic effect for each parameter and some two-parameter interaction effects. However, this approach cannot be applied if some of the parameters are categorical.

Taguchi's strategy is to avoid the need to allow for interactions at all. He recommends choosing the responses and control parameters carefully to avoid inducing spurious interactions (Phadke and Taguchi, 1987). This is, of course, always a good idea. But if interactions are inevitable for physical reasons, he recommends eliminating them by sliding the settings of the parameters involved. For example, in a photolithographic process, exposure time and aperture together determine the total amount of light energy

(Phadke *et al.*, 1983). Varying aperture and exposure time independently will probably result in an interaction. However, if exposure time settings are defined contingent on the settings of aperture, the degree of interaction can be reduced. We also recommend sliding settings if possible, but would still tend to allow for the interaction in the experiment plan.

Taguchi's strategy on interactions allows him to make frequent use of OA_{18}. This array can accommodate seven three-level parameters, but allows estimation of only one interaction (between the first and second columns). Each of the other interactions is partially confounded with several main effects (Sherry, 1987). Use of this array generally requires the assumption that all interactions besides the single one which can be estimated are negligible.

☐   Step 2B: construct noise plan

As mentioned earlier, the purpose of the noise plan is systematically to introduce 'real-world' variation in the experiment. Sometimes the most important sources of noise can be directly varied in the experiment. For example, the relative humidity in the customer's environment could have an important effect on the operation of a product. Since this humidity is not under the product designer's control, it is a noise variable. In a robust design experiment, however, the designer could easily test the product under different humidity conditions. So in this situation, humidity is a noise variable that can be directly introduced in the experiment.

The noise array is a plan for varying the levels of those noise variables that can be directly introduced in the experiment. Like the control array it can be constructed using an orthogonal array. Interactions between noise variables are generally not a concern, however, since the objective is to induce response variations and not to model the noise variables' effect in detail.

However, it is not always possible directly to introduce important noise variables in the experiment. This is often the case in manufacturing process design, where the major sources of noise tend to be variation of process parameter values from their nominal values, variation in raw materials and variation in the manufacturing environment. If a computer model for the process is available, many of these noises might be studied directly. But if no computer model is available, and physical experiments are being used, it is usually very difficult directly to vary the values of these noise variables. Instead, noise is introduced indirectly by taking replications of the responses in a way that is designed to capture the effects of these noise variables. For example, sources of variation in integrated circuit silicon wafer fabrication (Kacker and Shoemaker, 1986) include non-uniformity of temperature and gas composition inside a reactor. However, it is very difficult directly to vary these distributions by manipulating temperature and gas in different sections of the reactor. It is easier just to let the variations happen, and simply process wafers at a variety of positions in the reactor.

Using a noise array to introduce noise in an experiment can contribute greatly to the cost of the whole experiment, because the same noise array is followed for each control array row. Essentially, the combined control array × noise array experiment provides estimates of the interaction between every control parameter and every noise parameter. In addition, if interactions between control parameters were allowed for in the control array, then the combined array provides estimates of the higher-order interactions between those control parameters and each noise parameter.

This may be much more information than is needed to identify control parameters whose levels can be chosen to reduce the effects of the noise variables. An alternative is to combine the control parameters and noise variables in a single array, estimating only those interactions which are likely to exist. Shoemaker *et al.* (1988) discuss this and other approaches for studying noise more economically in robust design experiments.

Noise variables versus blocking variables

The concept of the noise variable might be confused with the related idea of the blocking variable explained by Fisher (1966) and Box *et al.* (1978, p. 102). Like noise variables, blocking variables arose from the recognition that in addition to controllable variables (that is, treatment factors in Fisher (1966)), there are other variables which affect the response. However, noise variables and blocking variables are used in different ways.

First of all, the objectives for introducing noise variables are different from those for introducing blocking variables. In robust design, a key objective is to reduce response variation by reducing the response's sensitivity to noise variables. If this objective is to be achieved the noise variables must interact with control parameters in their effect on the response. If this interaction exists, the control parameter levels can be chosen to reduce the effect of noise variables.

In contrast, the objective for introducing blocking variables is to remove their effect on the response so that the effect of the controllable variables can be evaluated. This is done by fitting a model which is additive in the blocking variables and may be additive or non-additive in the controllable variables. The assumption is that the blocking variables and the controllable variables do not interact.

■ Summary and discussion

The experiment planning methodology we have described is effective in a broad range of product and process design and improvement problems. It provides a single approach that can be followed when the problem includes qualitative parameters, quantitative parameters, or a mixture of both.

The methodology encourages the experimenter to study many parameters in a small experiment, allowing for expected interactions and expected curvature. The strategy carries the risk that large, unexpected interactions or unexpected curvature might be present, and could lead to inappropriate choices for new parameter levels. Because of this risk, the confirmation experiment that is the fourth step of the methodology carries considerable weight. It provides a direct comparison between performance at the initial and new parameter levels.

In certain special situations, however, other approaches may be more effective. For example, in problems calling for a small number of quantitative parameters, as is common in chemical processes, central composite designs (Box and Wilson, 1951) allow estimation of curvature for each parameter *and* interactions between parameters.

References

Addelman S. (1962). Orthogonal main-effect plans for asymmetrical factorial experiments. *Technometrics*, **4**(1), 21–46

American Supplier Institute (1984, 1985, 1986). *Proceedings of ASI Symposia on Taguchi Methods*, Volumes 1, 2 and 3, American Supplier Institute, Dearborn, MI

Box G. E. P., Hunter S. and Hunter W. G. (1978). *Statistics for Experimenters*. New York: Wiley

Box G. E. P. and Wilson K. B. (1951). On the experimental attainment of optimum conditions. *Journal of the Royal Statistical Society, Series B*, **13**(1), 1–38

Byrne D. M. and Taguchi S. (1986). The Taguchi approach to parameter design. *ASQC Quality Congress Transactions*, Anaheim, CA

Fisher R. A. (1966). *Design of Experiments*, 8th edn. Hafner (Macmillan)

Kacker R. N. (1985). Off-line quality control, parameter design and the Taguchi method, with discussion. *Journal of Quality Technology*, **17**(4), 176–209

Kacker R. N. (1986). Taguchi's quality philosophy: analysis and commentary. *Quality Progress*, December, 21–29

Kacker R. N. (1988). Taguchi's tables derived from Western published literature. In preparation

Kacker R. N. and Shoemaker A. C. (1986). Robust design: a cost-effective method for improving manufacturing processes. *AT&T Technical Journal*, **65**(2), 39–50

Kacker R. N. and Tsui K.-L. (1988). Interaction graphs. Submitted for publication

Lin K. M. and Kacker R. N. (1985). Optimizing the wave soldering process. *Electronic Packaging and Production*, February, 108–15

Pao T. W., Phadke M. S. and Sherrerd C. S. (1985). Computer response time optimization using orthogonal array experiments. In *Proceedings of IEEE International Conference on Communications*, IEEE Communications Society, June 1985, 890–5

Phadke M. S. (1982). Quality engineering using design of experiments. In *Proceedings of the American Statistical Association. Section on Statistical Education*. Cincinnati, Ohio, August 1982, 11–20

Phadke M. S. (1986). Design optimization case studies. *AT&T Technical Journal*, **65**(2), 51–68

Phadke M. S., Kackar R. N., Speeney D. V. and Grieco M. J. (1983). Off-line quality control in integrated circuit fabrication using experimental design. *The Bell System Technical Journal*, **62**(5), 1273–1309

Phadke M. S., Kacker R. N., Speeney D. V. and Grieco M. J. (1983) Off-line quality for robust design. *Conference Record, IEEE/IEICE Globecom 1987*, Tokyo, Japan, 1002–7

Raghavarao D. (1971). *Constructions and Combinatorial Problems in Design of Experiments*. New York: Wiley

Rao C. R. (1947). Factorial experiments derivable from combinatorial arrangements of arrays. *Journal of the Royal Statistical Society, Supplement*, **9**, 128–39

Sherry P. G. (1987). Confounding patterns in OA18. Unpublished manuscript

Shoemaker A. C., Tsui K.-L. and Wu C. F. J. (1988). Economical study of noise in robust design experiments. In preparation

Taguchi G. (1986). *Introduction to Quality Engineering: Designing Quality into Products and Processes*. Asian Productivity Organization, 4–14, Akasaka 8-chome, Minato-ku, Tokyo 107, Japan

Acknowledgements

In addition to the authors, those involved in developing this methodology include M. S. Phadke, J. Schofield, C. Sherrerd, P. G. Sherry and G. Ulrich, all of AT&T Bell Labs, and Prof. M. Steele of Princeton University. The authors gratefully acknowledge their considerable contributions, and in addition, those of Sig Amster, Ramon Leon and Kwok-Leung Tsui.

3.3

An integrated approach to quality engineering in support of design for manufacture

A. D. Brown,
P. R. Hale and
J. Parnaby

Lucas Industries plc, Birmingham

■ Introduction

The integration of quality and reliability engineering techniques in the product introduction and modification change control process is necessary for successfully satisfying customer requirements. This begins with the clear identification of true customer requirements, which are then ranked in order of importance and given visibility and traceability throughout the product introduction process. New, difficult to meet and important requirements are taken from one stage to the next with more detailed investigation being carried out at each subsequent stage. In this way all effort is focused and directed towards satisfying the customer. This is not only the next-in-line customer, for example original equipment manufacturer, but includes all those in the customer chain, which often amounts to a long list.

Discussions with customers when working towards defining their requirements must be pro-active, with all relevant options, opportunities and possibilities being offered, to ensure that the potential advantage given by innovative ideas is not lost.

This paper is reprinted by permission of the authors and the Council of the Institution of Mechanical Engineers from *Proceedings of the Institution of Mechanical Engineers* (1989), Part D, No. D1, pp. 55–63

An overview of how the techniques integrate within a total quality system is provided, together with cases drawn from the experience of Lucas Engineering and Systems Limited supporting Lucas operating companies in their quest to continuously improve. The case studies have been selected to illustrate the use of the techniques thought to be of greatest general interest. The interrelationships of the various techniques and their organizational and support needs in the engineering and manufacturing environment are given. A reference list is included for those interested in more detailed study.

■ The need for improvement

There should be little need now for the presentation of a case for improving product quality as a means of both protecting and gaining competitive advantage. The premise that 'quality improvement' is a must will be taken as read. However, it is worth just reflecting on both of the words 'quality' and 'improvement', before examining how to tackle the task.

Quality is defined as satisfying or exceeding customer expectations. Some people use different forms of words, but it is the meaning behind the words that is important; the goal is centred on expectations which is a higher target than conformance to specifications. Quality is used in this article in this higher sense; it has no direct connection to the man-in-the-street concept of degree of luxury or level of performance. It is diametrically opposed to any notions of acceptable levels of defects and presumes zero defects as the policy.

Improvement implies any necessary step changes that might be required to achieve best internationally demonstrated practice or to match customer perceptions of quality performance. It also implies embarking thereafter on a journey of continuous improvement in pursuit of rising customer expectations.

■ A total quality organization

Embarking on a total quality programme is a major change process for any organization which must in time reach all parts of the organization and result in changes in everyone's attitudes and behaviour. There is no panacea, no single-valued solution that will open the door to the secret garden. It requires a broad attack which will address issues of strategy, organization structure, systems and processes and culture as well as quality assurance procedures and quality techniques. The approach needs to be that of systems engineering, that is holistic; it needs to draw on general principles and put together the unique application of those principles to suit the particular circumstances prevailing. Tackled in this broad manner, a company can progress to being a

There are three functions that dictate how an organization functions and how that organization is perceived. These are shown below:

Total quality organization

Culture   Structure   Systems

Success comes from the right balance between these three factors.

*Culture: The combination of company values and management style and the employees' attitudes and reaction to these values supported by training.

*Structure: Formal reporting relationships usually shown by an organizational chart – trades off specialization and natural group integration of people, jobs and departments.

*Systems: Procedures formal or informal that match organizational 'ownerships', some of which may be supported by information technology.

Figure 1   The total quality organization.

total quality organization which satisfies the expectations of all stakeholders – customers, shareholders, employees and suppliers (Figure 1). The principle of a total quality approach is that of top quality performance in all functions in the organization such as design, manufacturing, materials supply, administration, etc., as a result of which high product quality is a by-product (Akao, 1986).

This paper focuses on quality engineering techniques, which on their own do not represent a complete quality system and certainly not a total quality programme. Techniques only achieve their potential when the company is organized and managed to make full and effective use of them, with leadership from the highest level. The emphasis must be on the definition of the basic business processes and their control.

A total quality organization is an organization that is totally committed to quality and one that:

Focuses on: Continuous process improvement
- Everything as a process
- The use of scientific methods – decisions supported by facts/data
- Perfection as the goal

Through:
- Universal participation
- Everyone
- Everywhere
- Individuals and teams

Resulting in: Customer satisfaction
  ● Exceeding expectations
For: Internal and external customers.

■ Quality engineering support techniques

Appendix 1 lists some basic problem-solving techniques. Techniques like statistical process control and quality circles are taught within the school curriculum in Japan! If employees at all levels are not comfortable with using such basic problem-solving techniques then there is little point in trying to install sophisticated methods.

Key off-line and on-line quality techniques lead to the heart of the issue, but before proceeding it is useful to be reminded of an application within the basic product introduction process which is illustrated in Figure 2 indicating the points at which off-line and on-line techniques can be applied. It should be noted that, as the figure implies, the effective way to handle the design engineering processes is very much one of parallel processing by a trained multi-disciplinary project team – simultaneous engineering as some

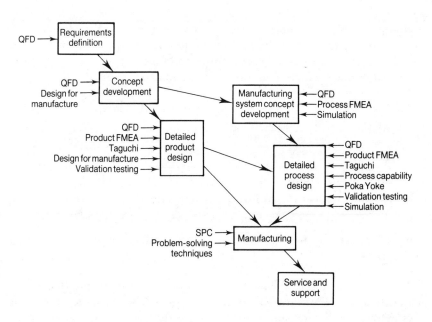

Figure 2   Quality techniques in the product introduction process.

call it. This ensures that all relevant skills are brought to bear on the problems as they are exposed.

The generally accepted reliability techniques listed in Figure 2, which form a backbone of any new product introduction, are described in detail in the references listed at the end of this paper. They must generally for effectiveness be applied by a mixed-discipline team as distinct from a single-subject specialist to ensure that all relevant ideas and experience are brought to bear on the problems. Also in Figure 2, when progressing from left to right, desk-based filtering processes eventually lead to a final subset of process or product problems which can only be solved experimentally so leading to application of Taguchi methods as the final step (including Shainin methods).

☐   **Quality function deployment**

The corner stone of sound quality foundations is QFD, quality function deployment. Figure 3 illustrates the principle whereby the specification or 'voice of the customer' is set down as customer attributes (inputs) down the left-hand side of the matrix with the resulting product design features and functions described along the top (outputs). A simple convention is then chosen to define strong, medium, weak (and, by omission, no) relationships. There are other conventions that choose more than three levels and distinguish between positive and negative relationships. You should choose your convention and stick to it. The customer attributes should include anything that is important to customer service, not just product perform-ance. Likewise the top of the matrix should include organizational issues where relevant as well as product features and functions.

Additions are then built into the basic matrix:

- Ranking relative importance to customers of the attributes.
- Mapping customer perceptions of a company's own products against competitor's.
- Mapping actual measurements of a company's own product perform-ance against competitor's.
- Setting targets for a company's own products.
- Listing a measure of difficulty and/or cost of achieving targets.
- Mapping the interrelations between product design features and functions (in the apex, above the matrix).

The trick then is to turn the product functions and features into the desired attributes (side of a new matrix) for the component characteristics (top of the new matrix). Full extension of this concept then allows the 'voice of the customer' to be cascaded down through the product introduction process via

	Angle of conveyor to be minimum	Good grip belt to be used	Determine minimum speed of conveyor	Ensure belt is good fit and aligned	Conveyors to plug into press socket	130-mm diameter wheels to be used	Non-fouling hood required on chute	Mount motor as low as possible	Use quick-release mechanisms
Ensure shoes do not fall back down conveyor	◉	●	◉	○					
Standard speed required on all conveyors			●	○					
Ensure shoes do not jam on conveyor	○	○	○	●			●		
Conveyor to be moved easily					●	●	○	◉	●
Stop shoes bouncing off conveyor on to floor	◉	◉	◉				●		
Conveyor to be stable	◉							●	
Points rating	10	13	16	11	9	9	19	12	9

○ Weak relationship
◉ Strong relationship
● Very strong relationship

Importance rating
○ = 1   ◉ = 3   ● = 9

Figure 3   The quality function deployment, QFD, method.

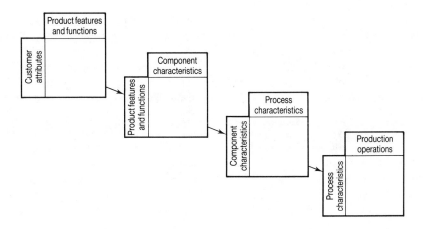

Figure 4   QFD – cascade.

process selection to operational instructions, as illustrated in Figure 4. At each level the matrix relates the important elements of 'how' to the important elements of what needs to be done. New, difficult to meet and important requirements are passed from one matrix to the next, thereby keeping effort correctly focused.

Lucas has used QFD in various ways. As well as working from the customer attributes for new product introductions Lucas has sometimes used the last steps, working from component characteristics down, to assist an investment programme for replacement of plant and equipment. It has also been used as a problem analysis tool, helping to solve an in-service problem which had defied earlier attempts to correct. A case study involving the use of QFD is detailed briefly in Appendix 2.

☐ Failure modes and effects analysis

Failure modes and effects analysis, FMEA, is now a very familiar tool which is extensively used in Lucas, both as design FMEA and process FMEA. QFD and FMEA are complementary as the first is targeted at satisfying customer expectations, the second at preventing failure to satisfy. However, as both are related and include detailed administrative tasks computer software has been developed to automate the administrative work, freeing engineers to do the thinking and implement the action plans.

FMEA is also used in a hierarchical manner where this simplifies its application. This applies particularly to complex systems with many components, where a 'bottom-up' approach would be too time consuming and costly. In this way the mundane design requirements do not receive the in–depth attention afforded to the critical characteristics.

Like the case for QFD, the important thing is to use the FMEA to assist quality planning. In this way effort is directed away from non-priority areas towards priority ones.

FMEA is also used for design of manufacturing systems and business systems as well as manufacturing processes. It is a very useful aid to designing, for example, reliable goods inwards systems and procedures, since once a process flowchart is drawn it is amenable to the application of FMEA to each element. A case study involving the use of FMEA is detailed in Appendix 2. The application of FMEA results in a basic set of decision trees summarizing quality hazards, and these can subsequently be used for supporting diagnostic processes and in expert systems, for example for process operation support.

☐ Taguchi methods

Quality loss

Taguchi's philosophies and methodologies are directed towards achieving highest quality at minimum cost. Quality, defined as satisfying or exceeding customer expectations or a failure to deliver it, is expressed in monetary terms as 'quality loss'. This recognizes that failure to meet customer expectations may result in direct cost to the customer and will certainly result in dissatisfaction, which in turn means less future sales, that is a reduction in market share, necessitating higher marketing and advertising costs, etc. All this is in addition to the direct supplier losses in scrap, rework, inspection, warranty, and so on, and is passed right down through the customer/supplier chain. In other words, the financial impact of poor quality is far greater than it appears on the surface.

Using QFD, customer requirements and expectations are turned into product design features and functions. This allows performance measures to be established and target performance levels to be set. The 'quality loss function' relates performance, in terms of deviations from target, to financial loss, demonstrating that the key to high quality is to minimize performance variability, which in turn minimizes quality loss.

As well as its philosophical implications in providing the justification for minimizing variability, the quality loss function has a number of direct applications. One area is in process control, which utilizes on-line feed-forward control techniques such as statistical process control, SPC. Another is in tolerance selection, where component tolerances are apportioned based on the required product tolerance, degree of influence of the component on product performance and cost implications.

Parameter design

The aspect of Taguchi methods that has received most attention is design optimization and, in particular, parameter design by experimentation. This focuses on the need to minimize performance variability as discussed earlier. Taguchi optimization methods have a role to play in both the design of the product and the design of the manufacturing systems and processes necessary to produce it. As with the techniques described earlier, the successful application of the methods is dependent on a multi–disciplinary approach and the matching of product design concepts and manufacturing strategies from as early a stage as possible in the development cycle.

The first step is concept development. More than one design concept should be considered and the pros and cons evaluated from customer, engineering and manufacturing perspectives. The next step is systems definition in which the concepts are refined into a selected product configuration which determines the nature of the product system, how its functions will be delivered, what materials will be used and what components will be required. Once the design progresses beyond this point any major changes will be very difficult, costly and time consuming to effect. The multi–disciplinary input here is at the heart of a right–first–time, preventative approach. Basic manufacturing options should be understood at this stage.

Parameter design is the determination of how parameter values affect the system performance and the selection of the values to be assigned to parameters. It determines which mix of parameters (dimensions, speeds, weights, spring rates, volumes, resistances, etc.) are employed to achieve the target system performance with greatest consistency. Using Taguchi methods encourages the selection of robust choices which tend to make the system performance independent of those factors that it is difficult, expensive or impossible to control. In other words, robustness is immunity to 'noises' experienced by the system (temperature, viscosity, vibration, wear, tolerances, voltage, age, etc.). From the types of parameters classified as 'noises' it is apparent that system reliability is implicit in the term 'robustness'. Experimental design and analysis techniques are used to identify critical design parameters, select optimum settings and provide a means of reaching a compromise design where there may be multiple, and in some cases conflicting, performance requirements.

While the use of parameter design is advocated as early as possible, it is recognized that in the concept development and systems definition stages it is not normally possible to conduct hardware experiments. However, it is still possible to take on board the Taguchi philosophy and consider the potential influence conceptually of both controllable parameters and 'noises' on system performance with the aid of cause-and-effect diagrams and input/output analysis (Appendix 1). FMEA used at a previous stage provides a useful input here. Simulation may be a substitute for prototyping, allowing experiments to be carried out on a model of the system. This is more generally applicable in the case of low-volume, high-cost products where the cost of hardware experimentation can be prohibitive at any stage. The type of modelling tools that may be used include mathematical models, finite element analysis and materials flow analysis used, for example, in tool design for plastics and rubber moulding and casting. Another area of considerable potential for optimization using a system model is that of electronic circuit design using, for example, the SPICE simulation package on a DAISY workstation.

Taguchi parameter design techniques are being used extensively for optimization of manufacturing processes. Typical areas of application include machining, welding, heat treatment, casting, plastics moulding, surface mount and flow soldering. This step should again be undertaken as early as possible to ensure that the selected manufacturing processes are capable of meeting and maintaining the required component performance and tolerances. Robustness of the process against variations in machines, operators and environmental conditions, for example, is as important as robustness of the product design. Reference should be made to the process FMEA and the combined results from these exercises used to determine what on-line controls will be needed.

Although the methods have been exploited in the optimization of processes, they have been used to a far lesser extent in manufacturing systems design. This is another aspect of design and development that can be tackled with the aid of simulation – in this case discrete-event simulation, which allows 'what–if' questions to be asked concerning the choice of manufacturing facilities and operating rules, with performance measured in terms of lead time, work in progress, etc. This should be matched against requirements and options considered in steps three and four of the QFD cascade (Figure 4).

Tolerance design

Tolerance design follows on from parameter design and is concerned with identifying and setting critical component tolerances. As distinct from tolerance selection, this is done by experimentation in a similar way to parameter design. Since Taguchi experiments are likely to consume significant resources, care should be taken not to use them indiscriminately as a problem-solving tool. FMEA can be used as a screen and QFD as a means of focusing on priorities. Costs of carrying out tolerance design experiments can be particularly high, unless simulation tools such as those mentioned earlier are used. This, coupled with the fact that they tend to result in recommendations for tighter tolerances and higher-grade materials for selected parameters, means that they are normally used when variability after minimization using parameter design is still higher than acceptable.

Case study: sensor design

One means of overcoming the pressures of short development time-scales for new product introductions is to carry out a development programme on a generic product family. By experimentation the influence of design parameters on product performance can be quantified. Referring to a database of the experimental results, parameter values can readily be set to achieve the desired product performance characteristics for a new customer requirement. Thus lead times are drastically reduced, since there is no longer the need for extensive prototyping on individual applications.

An example of this approach is the work carried out by Lucas Diesel Systems Limited on the design of variable inductance linear position sensors, of the type used in diesel fuel-injection equipment. These applications require high-performance, rugged sensors, able to operate with digital electronics and to withstand the hostile environment experienced by diesel engine mounted components.

A Taguchi experiment was carried out to:

- identify key design parameters;
- create a database of sensor performance against these parameters;
- extend design capability beyond the current design envelope.

The experiment involved the sensor and its interface circuit. The sensor, in its simplest form, consists of a coil wound on a bobbin, a static sleeve and a movable core which changes the inductance of the coil (Figure 5).

Figure 6 shows the nominal output characteristic of the sensor. This output, measured as a pulse width in microseconds for a given core displacement, must satisfy several requirements: low-base output, high sensitivity, good linearity and minimum susceptibility to external variations,

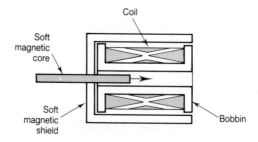

Figure 5   Basic mechanical arrangement of variable inductor.

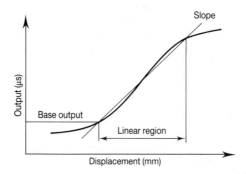

Figure 6   Nominal output characteristic of inductance sensor.

for example temperature and voltage. This last requirement of robustness against external 'noises' is particularly important, since to calibrate each sensor would be prohibitively expensive.

Eleven controllable factors, relating to the coil, sleeve, core, air gaps, interface circuit and external components, were studied in the experiment, each parameter being tested at one of three settings. In order to accommodate this number of variables the experiment was carried out according to, in Taguchi terms, an L27 orthogonal array, which necessitates 27 different builds of sensor. The core was displaced over 20 mm and readings of sensor output were taken at 2-mm intervals to produce an output curve for each of the 27 sensor designs. This procedure was repeated at four combinations of temperature and supply voltage, to ensure that the design selected was robust against variations in these uncontrollable 'noise' factors. For each of 108 test configurations, a graph of the 11 displacement readings was plotted to produce the S-shaped output characteristic curve, as shown in Figure 6. From these curves values of slope, linearity and base output were obtained.

Analysis of these measures enabled the identification of significant design features with respect to each aspect of sensor performance. The resulting database can be reassessed against customer requirements for future sensor applications, which should considerably reduce development time-scales, giving a more rapid response. Not only did the experiment fill the gaps between isolated pockets of design knowledge, it demonstrated that good designs can be achieved with settings beyond the ranges previously considered. This means far greater flexibility and scope for new designs. The results of the hardware testing can also be used to validate a computer model of the sensor, which can be used in the future to shorten testing time still further.

☐ **Process capability studies**

Process capability studies and statistical process control are now very familiar. It is not proposed to discuss them as techniques here, except to re-emphasize their importance. Unfamiliar readers are referred to vehicle builders' quality standards or the Society of Motor Manufacturers and Traders' reference document on statistical process control (1986).

The important point is what values of process capability, Cpk, are regarded as acceptable. The conventional standard is a Cpk of 1.33. However, best practice internationally is to push for continuous improvement, with values of 20 being claimed.

Lucas has applied the concept of process capability to robotic equipment by developing a test system that will actually determine the capability of multi-axis robots.

☐ **Poka Yoke**

In the drive towards zero defects, the use of Poka Yoke, or fool-proof devices, needs to be considered and implemented wherever possible. Such an approach can considerably simplify final acceptance testing (Shingo).

Three types of Poka Yoke may be identified:

(1) Contact type: This uses shape, dimensions or other physical properties of products to detect the contact or non-contact of a particular feature and hence prevent the manufacture of defects (for example, asymmetrical holes through which the components must pass).

(2) Constant number type: This detects errors if a fixed number of movements have not been made (for example, all parts have not been provided in a kit).

(3) Performance sequence type: This detects errors if the fixed steps in a sequence have not been performed or, alternatively, prevents incorrect operations from being performed, thus eliminating any defects.

There are two recognized ways in which a type of Poka Yoke may be activated:

(1)   Shut-out type: This prevents an incorrect action from taking place.
(2)   Attention type: This brings attention to an incorrect action but does not prevent its execution.

As the shut-out type of activation halts processing even if the operator is not paying attention, it is to be preferred (where possible) to attention-type activation where production continues if the operator does not notice the warning.

Recently 'fault tolerance' has emerged as a viable alternative to the two ways in which a type of Poka Yoke may be activated. Fault tolerance aims to make the manufacturing system robust enough to continue in the event of a fault occurring, without compromising product integrity, albeit with a probable reduced efficiency, until the fault is rectified. This can be extremely important with high-cost manufacturing systems, where minimum down-time is essential.

☐   **Quality plans**

Application of the off-line quality techniques makes the job of quality planning for the actual manufacturing operations a relatively straightforward, although detailed, task. The process capability studies and the FMEA, in particular, direct attention to where the control requirements are. It is important not to saturate the shop floor with control charts but to direct effort to priority areas. Figure 7 shows an input–output analysis of the quality planning task.

It should be noted that there is a feedback loop which should be used to adjust quality plans if necessary. This may mean increasing or decreasing the degree of control as progression is made along a path of continuous improvements or as learning occurs.

☐   **A reliability engineering programme**

Table 1 lists the main elements supporting a reliability policy and Table 2 summarizes the main tasks of a reliability programme.

☐   **On-line controls**

Statistical control of the process is, in the case of variables, able to prevent excursion outside the specification using feedforward principles. However, the second, very important, reason for having on-line control should not be

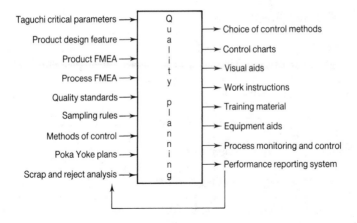

Figure 7   On-line quality control planning.

Table 1   The elements to be integrated under the stated reliability goal and policy.

1   A training programme for new product and process teams.
2   Design for manufacture, assembly and maintenance.
3   A reliability growth programme using Duane or some other relevant technique for mean–time–between–failure growth.
4   A fault categorization structure – by type.
5   An integrated information system – fault reporting, analysis and corrective action system, FRACAS.
6   A multi-disciplinary team approach.
7   An engineeering change control procedure supported by FRACAS.
8   A professional reliability engineering approach integrating design for function and manufacture with design and control of the manufacturing process.
9   A continuous improvement programme.

forgotten, namely, to assist in diagnosing the cause of the variation for the purpose of either eliminating the cause of variation or redesigning the product/process to make the product robust against the effects of the variation (Taguchi).

The philosophy must therefore be a thorough one of prevention in preference to detection, noting the following points:

(1)   Robustness is better than tighter tolerances.
(2)   Off-line preventative methods are better than on-line controls.
(3)   Eliminating the source of variation is better than controlling the variation.

Table 2 Reliability programme tasks.

1   Reliability programme plan/bar diagrams with targets.
2   Monitor/control of suppliers and a supplier development programme.
3   Programme reviews regularly by team leader.
4   Failure reporting, analysis and corrective action system.
5   A failure and change review board.
6   Environmental stress screening (full-searching automated testing).
7   Reliability growth programme with all relevant monitors.
8   Reliability qualification programme.
9   Production reliability acceptance test programme.
10 Manufacturing system design programme.
11 Process control and capability development programme plus Poka Yoke.
12 Communication and training programme.
13 Production test equipment standardization and simplification.
14 Assignment of work package ownership in the plan, that is a systems engineering approach.

(4)   Controlling the process is better than inspecting the process.

(5)   Inspecting the process is better than inspecting the product.

■ Conclusions

The need to use integrated quality techniques should be clear in pursuit of competitive business performance. For effectiveness of the use of the techniques it is essential to ensure that all necessary organizational support elements are in place. In particular it it important to ensure cross-functional and team or task-force-based applications since the narrow application of individual techniques by single specialist functions is very rarely satisfactory because of their very narrow view. A supporting framework such as BS5750/ISO9000/AN29000 can be a very beneficial element in an organization policy.

The techniques summarized are not substitutional; they form elements of a complementary group for application in series or parallel as a part of the total process of quality engineering.

References

Akao Y. (1986). Outlined description of quality development. *Standardisation and Quality Control*, Tokyo, April, 63–72

Akao Y., Harada A. and Matsumoto I. (1986). Quality and technology development. *Standardisation and Quality Control*, Tokyo, September, 57–70

Byrne D. M. and Taguchi S. (1986). The Taguchi approach to parameter design. Presented at *The International QC Forum*

Clausing D. P. (1986a). *Quality Function Deployment*. Mimeo Report, February

Clausing D. P. (1986b). *Quality Function Deployment*. Mimeo Report, September

Fortuna R. M. (1988). Beyond quality: taking SPC upstream. *Quality Prog.*, June, 23–8

Jun S. and Ryoji N. (1987). *Quality Deployment in Construction Industry*, February (Standardisation and Quality Control)

Kenny A. A. (1988). A new paradigm for quality assurance. *Quality Prog.*, June, 30–2

King R. (1987a). Better designs in half the time – implementing QFD in the USA. Presented at the *Growth Opportunities Alliance of Lawrence*, MA

King R. (1987b). Listening to the voice of the customer: using the quality function deployment system. *Nat. Productivity Rev.*, Summer, 227–81

Kogure M. and Akao Y. (1983). *Quality Function Deployment and CWQC in Japan*, October (Quality Press), 25–9

Parnaby J. (1987). Competitiveness via total quality of performance. *Progress in rubber and plastics technology*, **3**(1), 42–50

Proceedings of the Fourth Symposium on Taguchi Methods, October 1986 (American Supplier Institute Inc)

Proceedings of the Fifth Symposium on Taguchi Methods, October 1987 (American Supplier Institute Inc)

Ross P. J. (1988a). *Taguchi Techniques for Quality Engineering*. McGraw-Hill

Ross P. J. (1988b). The role of Taguchi methods and design of experiments in QFD. *Quality Prog.*, June, 41–7

Sato N. *Quality Function Expansion and Reliability*. Mimeo report

Shingo S. *Zero Quality Control: Source Inspection and Poka-Yoke System*. MA: Productivity Press

Society of Motor Manufacturers and Traders: Guidelines to Statistical Process Control (1986). London: Society of Motor Manufacturers and Traders

Sullivan L. P. (1986a). The seven stages in company-wide quality control. *Quality Prog.*, May, 77–83

Sullivan L. P. (1986b). Quality function deployment. *Quality Prog.*, June, 39–50

Sullivan L. P. (1987). The power of Taguchi methods. *Quality Prog.*, June, 76–9

Sullivan L. P. (1988). Policy management through quality function deployment. *Quality Prog.*, June, 18–20

Taguchi G. (1986). *Introduction to Quality Engineering – Designing Quality into Products and Processes*. New York: UNIPUB/Quality Resources

Takezawa N. and Takahashi M. (1986). Deployment of quality and reliability. *Standardisation and Quality Control*, Tokyo, October, 88–97

Vera D. D., Kenny A. A., Khan M. A. H. and Mayer M. (1988). An automotive case study. *Quality Prog.*, June, 35–8

■   Appendix 1

☐   Problem-solving techniques

Pareto analysis

Two-dimensional ranking by magnitude to determine the most important variables.

Cause-and-effect diagrams

Sometimes referred to as fishbone diagrams, these are a set of inputs associated with a particular output and relate causes and effects by exploding potential primary, secondary, etc. causes (inputs) from a measurable effect (output).

Histograms

Graphical ranking of variables.

Scattergrams

A pattern-recognition technique for exposing relationships between two variables.

Brainstorming

A structured approach to idea generation by teams which encourages idea generation and contribution from the team.

Input/output analysis

The basic systems engineering technique which views things as a process with defined boundaries which produce outputs and require inputs to enable the process to work. The inputs must have a supplier and the outputs must have a customer process who finds them necessary.

Problem selection

A simple technique for allowing teams to rank and select problems to work on.

■ **Appendix 2**

☐ **Quality function deployment, QFD, and failure modes and effect analysis, FMEA, case study**

Recently QFD and FMEA were employed to ensure that a particularly troublesome oil leakage problem would be eradicated by the introduction of a new flow control valve, without any new problems being generated.

The unit in question was a fuel flow regulator for an aerospace application. A shaft was rotating at approximately 1500 rpm. High-pressure fuel at 1200 lbf/in^2 was supplied to one side of a bearing. This reduced to approximately 150 lbf/in^2 across the bearing and was maintained by a backing pump. A garter seal, running on the shaft, retained the 150 lbf/in^2 and prevented the fuel draining into a tank at atmospheric pressure.

Various attempts had been made to eliminate the leak by changing the shaft and seal material and improving the shaft surface finish. The minimal improvement that resulted from this approach led to the decision to investigate the option of introducing a new unit.

The principle of operation of the proposed new unit was similar to one currently in production, with a reasonable reliability record. The detailed design of the unit did, however, differ in that the replacement unit had to be of the same external dimensions as the unit currently fitted. It was decided that the easiest and most cost-effective way of satisfying this requirement was to utilize the existing housings, machined and sleeved to accept new internal components. This resulted in the obvious concern that although the leakage problem would be solved others would be generated, and these were systematically examined.

The proposed unit was servo assisted, needing a smaller volume of air for operation than the non-servo-assisted unit being replaced. This resulted in potential performance differences which required investigation and customer consultation to determine the seriousness of the effect on the aircraft. Investigation revealed the not-surprising result that the servo unit responded much faster to cockpit input signals than a non-servo version. This was of concern to the customer because it would be possible for an aircraft to be fitted with both versions of the unit. If this occurred on opposite wings the aircraft would skew when the throttle levers were moved through the same angle. In order to minimize this effect the volumes of air used in the servo unit would be matched to that of the current unit.

An FMEA was produced, based on functionally critical items identified in the QFD matrices. Both of these documents were used to form the basis of design reviews involving multi-disciplinary teams including customer representation. In this way the potential sources of problems were examined and actions identified to reduce the risk of undetected failure occurring, or reduce the severity of the failure should it occur. This was accomplished by detailing the deviation from the functionally critical items, determining the effect of that deviation and rating the severity of the effect either in house or on the customer. Knowledge of the manufacturing system used enabled the causes of the deviation to be identified and the likelihood of the causes occurring to be rated. Additionally the likelihood of detecting any such failure was assessed and rated from a knowledge of the on-line control procedures.

Prior to production of the FMEA, two QFD matrices were used to translate the customer requirements, generated by a Lucas team and discussed with the customer at the design reviews, to product requirements and then to component requirements. A cross-functional team of four Lucas personnel were involved in the production of the QFD matrices and FMEA. At each translation only the characteristics strongly related to the customer or product requirements were carried forward for further investigation. In that way attention was focused, allowing resources to be directed in the most beneficial areas.

The result of the exercise was a unit designed professionally and systematically to eliminate problems and more likely to satisfy customer requirements and hence provide satisfaction. Some readers may say this would be achieved by systems and 'good' practices already in place in their own companies; for a time the authors also believed they were doing all that they could only to find that they were not!

3.4
Product design for manufacture and assembly

G. Boothroyd and
P. Dewhurst
University of Rhode Island

Design for manufacture, DFM, means different things to different people. For the individual whose task is to consider the design of a single component, DFM means the avoidance of component features that are unnecessarily expensive to produce. Examples include the following:

- Specification of surfaces that are smoother than necessary on a machined component, necessitating additional finishing operations.
- Specification of wide variations in the wall thickness of an injection-moulded component.
- Specification of too-small fillet radii in a forged component.
- Specification of internal apertures too close to the bend line of a sheet-metal component.

Alternatively, the DFM of a single component might involve minimizing material costs or making the optimum choice of materials and processes to achieve a particular result. For example, can the component be cold-headed and finish-machined rather than machined from bar stock? All of these considerations are important and can affect the cost of manufacture. They represent only the fine-tuning of costs, however, and by the time such considerations are made, the opportunities for significant savings may have been lost.

This paper is reprinted by permission of the authors and the Society of Manufacturing Engineers from *Manufacturing Engineering*, April 1988, pp. 42–6

It is important to differentiate between component or part DFM and product DFM. The former represents only the fine-tuning process undertaken once the product form has been decided upon; the latter attacks the fundamental problem of the effect of product structure on total manufacturing costs.

The key to successful product DFM is product simplification through design for assembly, DFA. DFA techniques primarily aim to simplify the product structure so that assembly costs are reduced. Experience is showing, however, that the consequent reductions in part costs often far outweigh the assembly cost reductions. Even more important, the elimination of parts as a result of DFA has several secondary benefits more difficult to quantify, such as improved reliability and reduction in inventory and production control costs. DFA, therefore, means much more than design to reduce assembly costs and, in fact, is central to the issue of product DFM. In other words, part DFM is the icing on the cake; product DFM through DFA *is* the cake.

DFA derives its name from a recognition of the need to consider assembly problems at the early stages of design; it therefore entails the analysis of both product and part design. For some years now, an assembly evaluation method, AEM, has been in use at Hitachi. In this proprietary method, commonly referred to as the Hitachi method (Miyakawa and Ohashi, 1986), assembly element symbols are selected from a small array of possible choices. Combinations of the symbols then represent the complete assembly operation for a particular part. Penalty points associated with each symbol are substituted into an equation, resulting in a numerical rating for the design. The higher the rating, the better the design.

Another quantitative method, developed by the authors and known as the Boothroyd and Dewhurst method (1987a), involves two principal steps:

- The application of criteria to each part to determine whether, theoretically, it should be separate from all the other parts in the assembly.

- An estimate of the handling and assembly costs for each part using the appropriate assembly process – manual, robotic or high-speed automatic.

The first step, which involves minimizing the parts count, is the most important. It guides the designer towards the kind of product simplification that can result in substantial savings in product costs. It also provides a basis for measuring the quality of a design from an assembly standpoint. During the second step, cost figures are generated that allow the designer to judge whether suggested design changes will result in meaningful savings in assembly cost.

The third quantitative method used in industry is the GE/Hitachi method (Hock, 1987), which is basically the Hitachi method with the Boothroyd and Dewhurst criteria for parts-count reduction added.

For business reasons, companies are seldom prepared to release their manufacturing cost information. One reason is that many companies are not sufficiently confident about their costing procedures to want manufacturing costs made public for general discussion. In such an environment, designers will often not be informed of the cost of manufacturing the product they have been designing. Moreover, designers do not usually have the tools necessary to obtain immediate cost estimates relating to alternative product design schemes. Typically, a product will have been designed and detailed and a prototype manufactured before a manufacturing cost estimate is attempted. Unfortunately, by then it is too late. The opportunity to consider radically different product structures has been lost, and among those design alternatives might have been a version that is substantially less expensive to produce.

Currently, there is much interest in having product DFM and DFA techniques available on CAD/CAM systems. By the time a proposed product design has been sufficiently detailed to enter it into the CAD/CAM system, however, it is already too late to make radical changes. A CAD representation of a new product is an excellent vehicle for making effortless detail changes, such as moving holes and changing draft angles. But for considering product structure alternatives, such as the choice of several machined parts versus one die casting, a CAD system is not nearly as useful. These basic, fundamentally important decisions must be made at the early sketch stage in product design.

A conflict thus exists. On the one hand, the designer needs cost estimates as a basis for making sound decisions; on the other hand, the product design is not sufficiently firm to allow estimates to be made using currently available techniques. The means of overcoming this dilemma is another key to successful product DFM – namely early cost estimating.

Many individuals are proposing rule-based or axiomatic methods to help designers achieve efficient designs. The axiomatic approach proposed by N. P. Suh and his co-workers (1978) was based on attempts to identify common properties of successful designs. These common properties, such as how the proposed design satisfies the functional requirements, were then proposed as axioms of good design. Design axioms can thus be viewed as global product guidelines that can co-exist with component guidelines for details such as hole spacings, fillet radii and draft angles. However, axiomatic approaches have two major weaknesses when manufacturing is considered in the early stages of product design. Both of these weaknesses are directly related to cost.

First, the axiomatic approach does not provide any means of making judgements between the centrally important trade-offs posed by possible alternative choices of different materials and processes. Second, at the detail level, guidelines tend to lead designers in an essentially fruitless direction. This is because manufacturing guidelines are invariably intended to make individual processing steps as efficient as possible. Following such guidelines might lead to the design of all of the bend lines of a sheet-metal part in a single plane, the avoidance of side holes or depressions in moulded parts, the minimization of the number of steps in a part to be made by powder metallurgy, and so on. With this approach, the tendency is to design relatively simple individual components, which will invariably lead to high total fabrication and assembly costs.

A DFM system must therefore be able to predict both assembly costs and component manufacturing costs at the earliest stages of product design. Only in this way will it be possible to design a product that takes maximum advantage of the capabilities of chosen manufacturing processes within the constraints imposed by anticipated production volumes. In many situations, this will simply mean providing designers or design teams with the software tools that will enable them to make sound judgements from a range of choices. These choices may involve designs necessitating increased tooling costs but fewer different parts and reduced assembly costs.

It is anticipated that the product DFM considerations will always start with DFA. To aid designers in implementing these techniques, the authors have developed DFA software (Boothroyd and Dewhurst, 1987b) that allows a designer to establish an efficient assembly sequence for a proposed new product concept. The software then questions the relationship between the parts and gives an assembly efficiency rating, together with estimated assembly costs.

The DFA process uses the assembly sequence as a vehicle for analysing the product structure in order to force the design towards more integrated solutions with a reduced parts count. This result of DFA is often the most important one in achieving total product cost reductions. Thus, DFA analyses must be supported by techniques that will allow the design team to make early estimates of material, processing and tooling costs. Only in this way can different designs, with different numbers of parts and perhaps using different materials and processes, be compared before a detailed design commitment is made.

To illustrate this approach, consider the simple spindle/housing assembly shown in Figure 1. A DFA software analysis of the proposed sheet-metal housing, two nylon bearing inserts, six screws and a machined steel spindle gives an assembly efficiency rating of only 7%. This low rating arises mainly from the fact that there are 10 separate parts in the assembly, whereas theoretically only two should be necessary.

Application of the DFA minimum parts count criteria reveals that the housing and shaft must be separate because of their relative motion. None of the remaining parts satisfies the criteria, however, because they do not move relative to the housing; they can all be of the same material and do not have to be separate in order to assemble the spindle into the housing. These considerations lead the designer to consider a two-part design, and assuming that the bearing surfaces should be of nylon, the alternative design shown in Figure 2 might be proposed. In this case, the spindle is unchanged, but the

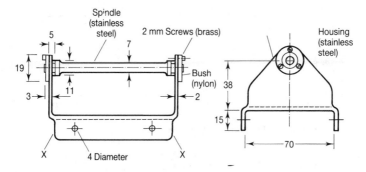

Figure 1 This spindle/housing assembly (sheet metal housing) has 10 separate parts and an assembly efficiency rating of 7%.

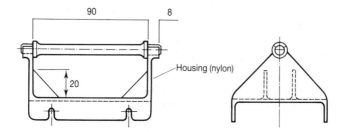

Figure 2 The two-part design, utilizing an injection-moulded nylon housing, has an assembly efficiency rating of 93%.

remaining parts have been combined into one injection–moulded nylon housing with an assembly efficiency of 93%.

Using the authors' techniques for early cost estimating (Dewhurst and Boothroyd, 1987), the cost breakdowns for the various parts used in the two designs were obtained (Figure 3). It can be seen that DFA has resulted in significant savings in assembly, material, manufacturing and tooling costs. Even though this was a DFA study, the largest savings are in material costs, which have been reduced from $2.73 to only $0.40 – principally by changing the material for the housing and eliminating the screws. The savings in manufacturing costs can be mainly attributed to the elimination of all the drilling and tapping operations, which cost a total of $1.35. This cost, together with the cost of the screws themselves, amounts to $2.07 and accounts for 34% of the original design's cost. These figures would indicate to the designer the desirability of seeking alternative methods of securing the bushes in the housing if the design with combined housing and bushes was not considered practicable.

A. Design using sheet metal housing

	Assembly	Material	Manufacturing	Tooling
Housing	0.02	1.74	1.56[1]	7.830[3]
Bush (2)	0.09	0.01	0.06[2]	9.030[4]
Screw (6)	0.35	0.72	–	–
Spindle	0.04	0.26	1.29	–
Total	0.50	2.73	2.91	16.860

[1] Includes $1.35 for drilling and tapping screw holes
[2] Moulded bushings have three-cored holes for screw clearance
[3] Three separate die sets for blanking, punching and bending
[4] Ten-cavity mould for least-cost manufacture

B. Design using injection-moulded housing

	Assembly	Material	Manufacturing	Tooling
Housing	0.02	0.14	0.24	10.050[1]
Spindle	0.02	0.26	1.29	–
Total	0.04	0.40	1.53	10.050

[1] Two-cavity mould for least-cost manufacture

Figure 3 A comparison of the two spindle/housing assembly designs shows significant cost reductions as a benefit of DFA.

The results of this DFA analysis, combined with early cost-estimating methods, illustrate the kind of result that can be obtained by using DFA as the first step in a product DFM study. Of course, it is also possible to achieve savings by considering changes in the product design that are directed at reductions in individual part costs.

For example, the rounded corners in the sheet metal housing (marked X in Figure 1) contribute unnecessarily to material and tooling costs. If these corners were squared, the allowance in the strip width of 4 mm per side would not be necessary, thereby resulting in material cost savings of 8%, or $0.14 per part. Additionally, a part-off die would be employed instead of a blanking die, resulting in die cost savings of around $1000. A design change that would reduce manufacturing costs would be the use of self-tapping screws and pierced holes in the sheet-metal part to avoid the drilling and tapping operations costing $1.35.

Looking at the costs of the proposed design (Figure 3(B)), it is clear that attention should be paid to the design of the spindle in order to reduce material and manufacturing costs. In this case, if the two larger diameter features could be eliminated so that the spindle could be machined from smaller diameter stock with reduced bearing surface diameters at each end, materials costs would be reduced from $0.26 to $0.11 and manufacturing costs from $1.29 to $0.85, a total savings of $0.44. Although these

Estimation of injection moulding costs for: housing Thermoplastic: 6/6 nylon	
Dimensional data	**Part complexity**
Part volume = 25.51 cm^3	Outer surface or cavity (0.5)? 2
Projected area = 71.25 cm^2	Inner surface or core (0.5)? 0
L = 95 mm W = 75 mm D = 57 mm	**Mould complexity**
Thickness: Maximum = 2.5 mm	
Average = 2 mm	Standard two-plate mould? Y
Quality and appearance	Three-plate mould? N
Tolerance factor (0.5)? 3	Multi-plate stacked mould ? N
Appearance factor (0.5)? 1	Hot runner system? N
Coloured resin? N	Number of side cores or pulls? 0
Textured surface? N	Number of unscrewing devices? 0
At any time press: <H>ELP, <V>OLUME, <A>REA, OR <C>OMPLEXITY CALCULATOR	

Figure 4   Shown are the inputs required to estimate the costs of injection moulding the housing.

considerations are important, they are only the tip of the iceberg compared with the results of DFA analysis.

The early cost-estimating programs that were used to determine the part costs given in Figure 3 have been developed for use with microcomputers and are based on the results of an extensive research program at the University of Rhode Island. These programs are designed so that they can be easily applied at the sketch stage before detail drawings are available and before precise processing and tooling specifications have been made.

For example, Figure 4 shows the inputs that are required to estimate the costs of injection moulding the housing shown in Figure 2. It can be seen that Nylon-6/6 has been selected. This has been obtained from a database that contains material costs and moulding parameters.

The footnote in Figure 4 indicates that support calculators are available for determining the correct responses for the housing's volume, projected area and geometric complexity. Figure 5 gives the resulting cost breakdown for the housing. Working together with an injection-moulding machine database, the program first selects the optimum number of cavities from the trade-off between cavity manufacturing costs, machine rates and the number of parts produced for each machine cycle. The program then produces a breakdown of mould, processing and material costs. These are given in the right-hand column of Figure 5 and are added to produce the part-cost estimate in the lower right-hand corner.

In a similar manner, the machining program works with two user-editable databases, one for machine tools and one for materials. The user selects the material, specifies the workpiece dimensions, selects the machine tool and specifies the cutting operation. The program then estimates the

Estimated injection moulding costs for: housing			Thermoplastic: 6/6 nylon		
Total production volume (thousands)	Number of cavities	Total mould base costs ($)	Cavity/core manufacturing costs ($)	Total mould cost ($)	Mould cost per cent (cents)
100	2	3589	6462	10.051	10.1

Select required option:
1. Screen edit
2. Show mould cost/cycle elements
3. Print results and responses
4. Change basic cost data
5. Change responses/polymer
6. Exit

Machine size (kN)	Machine rate ($/h)	Cycle time (seconds)	Manufacturing cost per part (cents)
1600	72	20.3	23.9

Part volume (cm³)	Part weight (grams)	Polymer cost ($/kg)	Polymer cost per cent (cents)
26	29	4.69	13.5

Total part cost (cents) = 47.5

Figure 5   The cost breakdown for the proposed injection-moulded nylon housing. The total part cost appears in the lower right-hand corner.

Material: Stainless steel – ferritic-free machining

Spindle   Machine: Manual turret lathe

Operation number	Operation <S>TEEL <D>ISPOSABLE <C>ARBIDE <G>RIND <R>OUGH OR <F>INISH		Load/unload time, etc. (seconds) Set-up time (h)	Set tool, engage cut, etc. (seconds) Number of operations	Volume (in³)	Area (in²)	Machining time (seconds)	Operation cost ($)
1 Face	F C		1.41	29	1	9	–	0.15	0.66	0.30
2 Cylindrical turning	R C		0.22	–	1	9	0.03	0.28	2	0.28
3 Cylindrical turning	F C		–	–	1	9	–	0.28	0.50	0.07
4 Cylindrical turning	R C		–	–	1	9	0.28	2.98	16	0.19
5 Cylindrical turning	F C		–	–	1	9	–	2.98	4	0.10

Batch > 10 000   Totals   3.48   58   81   0.35   7.50   27   1.29

Cost/part ($):   Material = 0.26   Set-up = 0.01   Operations = 1.29   Total = 1.56

(Move indicator to required row/column/page using keypad functions)
Press <INS>ERT, <DEL>ETE, <C>HANGE, <M>ATERIAL, <H>ELP, or <O>K

Figure 6   Results from entering the machining operations for the spindle appear on the computerized worksheet.

machining costs for the operation. As each machining operation is entered, the results appear on a worksheet where allowances are automatically added for non-productive time and tool replacement costs. Figure 6 illustrates a portion of the worksheet that results from entering the machining operations for the spindle.

The techniques of DFA and DFM can play a major role in reducing costs and increasing productivity. Recognition of this fact is also increasing the demand for cost-estimating tools that allow design teams to make the necessary trade-offs at the early concept stages of design. These techniques and tools can play a significant part in helping US industry regain its competitive edge in world markets.

References

Boothroyd G. and Dewhurst P. (1987a). *Product Design for Assembly Handbook.* Wakefield, RI: Boothroyd Dewhurst, Inc

Boothroyd G. and Dewhurst P. (1987b). *The Design for Assembly Software Toolkit.* Wakefield, RI: Boothroyd Dewhurst, Inc

Dewhurst P. and Boothroyd, G. (1987). Early cost estimating in product design. In: *Proceedings, Second International Conference on Product Design for Manufacture and Assembly*, April

Hock G. (1987). Designing for productivity. *Target*, Summer, 14

Miyakawa S. and Ohashi T. (1986). The Hitachi assemblability evaluation method (AEM). In *Proceedings, First International Conference on Product Design for Assembly*, April

Suh N. P., Bell A. C. and Gossard, D. C. (1978). On an axiomatic approach to manufacturing and manufacturing systems. *ASME Journal of Engineering for Industry*, **100**(2)

3.5
Design for assembly of a direct current motor

H. A. ElMaraghy, L. Knoll and B. Johns
McMaster University

M. Pavlik
Motor Division of Electrohome, Canada

■ Introduction

To decrease the cost of manufacturing a product, improving the efficiency of the manufacturing process is one solution; however, even efficient processes may not produce a cost–competitive product. The true problem may be rooted at the first step in manufacturing: the design of the product may make it difficult to produce. In an effort to improve producibility, design for assembly procedures and guidelines have been developed to aid in the identification of design problems and to make the designer conscious of how the design affects the process. The underlying principles of most design for assembly procedures is to reduce the number of parts in an assembly and to make those remaining parts easy to handle and assemble.

This paper discusses the design for assembly of a family of fractional-horsepower DC motors. There is an increasing demand for DC motors in the automotive industry today and therefore customer specifications will ultimately be one of the largest considerations in the motor redesign.

This paper is reprinted by permission of the authors and IFS Ltd from *Developments in Assembly Automation*, March 1988, pp. 191–200

An existing production fractional-horsepower DC motor (hereafter referred to as the 'old DC motor') is the candidate for redesign to facilitate assembly. This motor had been previously assembled manually and both the number of parts and the assembly cost warranted the investigation of a redesign. Also, to be compatible with the automotive market, a radio frequency interference, RFI, filter has been incorporated into the design, and material and motor construction have been considered by the manufacturer to ensure compliance with automotive specifications.

The new motor design selected for implementation is based on both the feasibility of manual and automatic assembly. During initial production, the assembly system currently consists of manual workstations with the introduction of a robotic workstation, for automatic final motor assembly, which will be gradually phased in through cost justification in the future.

■ The old motor design

It is important to differentiate between fabrication and assembly for the purpose of an assembly analysis. Fabricated parts are defined as those parts which cannot be disassembled to produce two discrete parts. Subassemblies are composed of fabricated parts and other subassemblies. The old DC motor consists of five main subassemblies as well as the required fasteners as shown in Figure 1. The motor assembly therefore, is analysed according to the following tasks:

(1) Opposite-commutator (opposite-comm) endcap assembly (without brushplate).

(2) Comm endcap subassembly (with brushplate).

(3) Magnet/flux ring subassembly.

(4) Brushplate subassembly.

(5) Armature subassembly.

(6) Final assembly.

□ Armature subassembly

The armature laminations, fibres, wires, etc., are manufactured using hard automation and will be considered as a single unit for the DFA analysis. The parts which comprise the armature are stacked along the principal axis and a build-up in tolerances occurs which affects the bearing-to-bearing distance. Therefore, during the final motor assembly process, the bearing-to-bearing distance is set by adding nylatron washers of varying thicknesses to each end of the armature shaft. It is the addition of these washers and the oil throw

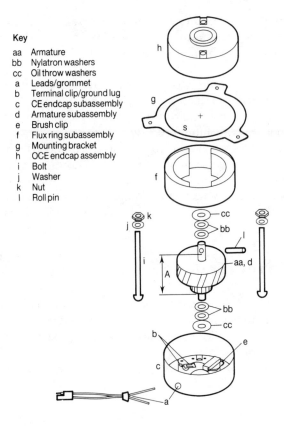

Key

aa	Armature
bb	Nylatron washers
cc	Oil throw washers
a	Leads/grommet
b	Terminal clip/ground lug
c	CE endcap subassembly
d	Armature subassembly
e	Brush clip
f	Flux ring subassembly
g	Mounting bracket
h	OCE endcap assembly
i	Bolt
j	Washer
k	Nut
l	Roll pin

Figure 1 Old motor assembly.

washers, together with the armature, which will be considered to be the armature subassembly task.

☐ Opposite-comm-end, OCE, endcap assembly (without brushplate)

The following is a parts list for the old OCE endcap subassembly (Figure 2):

(d)   four rivets

(e)   oil catcher

(f)   gasket

(g)   oiled felt

(h)   metal spring

(i)   spherical bearing

(j)   OCE endcap.

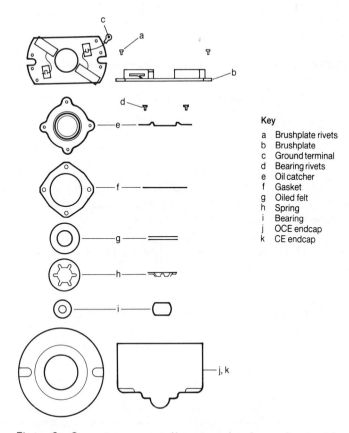

Key
a   Brushplate rivets
b   Brushplate
c   Ground terminal
d   Bearing rivets
e   Oil catcher
f   Gasket
g   Oiled felt
h   Spring
i   Bearing
j   OCE endcap
k   CE endcap

Figure 2   Opposite-comm-end/comm-end endcap subassemblies.

The oil catcher is positioned on the riveting fixture by locating pins through the rivet holes. Next, the gasket, oiled felt, spring and bearing are placed on the oil catcher. The endcap is then inverted and the rivet holes are located on the same pins as the oil catcher. An automatic riveting process feeds and rivets the assembly in one stroke. Both felt and bearing are lubricated before assembly.

☐   **Comm-end, CE, endcap (with brushplate)**

The parts in this subassembly are identical to the previous subassembly with the exception that the comm-end endcap has slightly different features and the brushplate is added (Figure 2):

(a)   brushplate and ground terminal (c) subassembly
(b)   four rivets (for brushplate)

(d)   four rivets

(e)   oil catcher

(f)   gasket

(g)   oiled felt

(h)   metal spring

(i)   spherical bearing

(k)   CE endcap.

The assembly process is also identical to the opposite-comm-end endcap assembly with the exception of the brushplate subassembly which is riveted at another machine. A ground terminal from one of the brushes is placed over one of the four locating pins. The brushplate itself and the comm-end endcap is then placed over the same pins. All four rivets are automatically inserted and fastened in a single stroke. The rivet makes the electrical connection from the ground terminal to the motor casing.

☐   Magnet and flux ring subassembly

(a)   magnet clamp

(b)   two permanent magnets

(c)   steel flux ring.

Prior to assembly, the flux ring must be phosphate coated. The magnets are fastened to the flux ring by an epoxy adhesive. A needle-nosed pliers is used to squeeze the spring clamp and insert it into position inside the flux ring. When released, it securely holds the magnets in position as the glue cures. When partially cured, the magnet clamp is removed and the motor is assembled. Then, the flux density is checked, and the glue is left to cure for another eight hours.

☐   Brushplate subassembly

Referring to Figure 3, the following parts are assembled together to form the brushplate subassembly:

(a)   ground terminal

(b)   brushplate

(c)   two brush tubes

(d)   two terminal clips

(e)   two springs

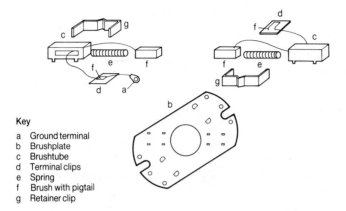

Key

a Ground terminal
b Brushplate
c Brushtube
d Terminal clips
e Spring
f Brush with pigtail
g Retainer clip

Figure 3 Brushplate subassembly.

(f)   two brushes with pigtails

(g)   two retainer clips.

The brushplate is aligned in a jig and a brush is positioned so that the pigtail extends through the slot in the brush tube. The spring is then placed in the brush tube and the brush is pushed in the tube, compressing the spring. A retainer clip is pressed on to the tube to hold the brush inside the tube. The pigtail is then hooked under the terminal clip and the process is repeated for the second brush. In a previous operation, a ground terminal has been crimped on one pigtail.

☐ **Final assembly**

During the final assembly process, the following subassemblies and parts are assembled:

(aa)   armature

(bb)   nylatron washers

(cc)   oil throw washers

(dd)   nylatron washers

(a)   leads/strain relief grommet

(c)   CE endcap

(d)   armature subassembly

(e)   brush clip

(f)   flux ring

(g)   mounting bracket

(h)   OCE endcap

(i)   two bolts

(j)   two washers

(k)   two nuts

(l)   roll pin.

The final assembly process requires four stations and seven workers. At the first station, the power leads are fed through the hole in the comm–end endcap and soldered to the brushplate. The assembled endcaps are then passed, by conveyor, to the next station where they are placed in a fixture. Washers of varying thicknesses are added to both ends of the armature to obtain the precise bearing-to-bearing dimension 'A' (Figure 1). The armature is removed from the jig and oil throw washers are fitted to each end of the shaft. The comm–end shaft is inserted into the comm–end endcap bearing and the retainer clips on the brush tubes are removed.

At the next station, the flux ring is inserted around the armature and into the comm–end endcap and the mounting bracket is added. The mounting bracket is added with the side stamped 'S' facing the shaft end. The opposite comm–end endcap is fitted to the armature and bolts are inserted from the comm–end endcap side. Nuts and washers are added and the nuts are torqued down. Thereafter, the finished assembly moves to the final station.

The assembly undergoes inspection at the last station. Endplay is checked and changed if required and the motor is set aside and 'run–in'. After the roll pin is inserted in the shaft, the motor goes to the test station where torque, speed and current are measured using a computer. If it passes the tests, the motor is packed and prepared for shipping. Should the motor not have sufficient torque or speed, it is returned for remagnetization.

The total assembly process also includes non–value–added operations such as refilling boxes of parts and moving the subassemblies and parts on the shop floor. By reducing the number of parts, inventories, parts flows and other industrial engineering operations will be simplified.

■ Old DC motor design analysis

The evaluation of the old DC motor for ease of manual assembly and design efficiency was done using a design for assembly, DFA, method developed by Boothroyd and Dewhurst (1983). This method involves estimating the manual handling time for each part in the assembly and estimating the manual assembly time for each operation in the total assembly based on the part design and assembly characteristics. The manual handling time is based on the symmetry and size of the parts as well as the ease with which the parts can be grasped and manipulated. Considerations are given to such part characteristics as flexibility, special tools or the requirement of two hands for manipulation. The operation time is based on how easily the parts can reach their desired location, the fits involved, the requirement of special tools and the number of hands required for the job. Consideration is also given to operations such as gluing and welding. It should be noted that the times obtained from the charts should not be interpreted absolutely. These times are based on general design attributes and should serve only as a relative indicator when comparing two designs.

The result of the analysis is a design efficiency, which is defined as the ratio of the 'theoretical time' for an assembly containing a 'theoretical' number of parts to the total assembly time derived from the DFA analysis of the actual product:

$$\text{Design efficiency} = \frac{\text{theoretical assembly time}}{\text{DFA assembly time}}$$

The theoretical time is calculated as the time taken to assemble a 'theoretical' part (three seconds) times the number of 'theoretical' parts in the assembly. The theoretical number of parts is determined by asking three questions:

(1) Does this part move relative to another?
(2) Do the mating parts have to be made of different materials?
(3) Do the parts have to be separate to allow servicing before/after assembly?

If the answer to any of these questions is 'yes', then that part cannot be eliminated. An additional question is asked if there is a repetition of the assembly:

(4) To how many of these parts does this apply?

The numerical answer to this question is the theoretical number of parts. Therefore, the formula for design efficiency is:

$$\text{Design efficiency} = \frac{3 \times \text{theoretical minimum number of parts}}{\text{DFA assembly time}}$$

It should be noted that even the best redesigns do not always reduce the number of parts to the minimum theoretical number for various reasons.

The motor was taken apart and analysed using these manual handling and manual insertion considerations. The overall characteristics of the old motor design are as follows:

- Total estimated assembly time:
- Overall design efficiency:
- Best design efficiency:
- Worst design efficiency:
- Total number of parts:

424.4 seconds.
14.8%.
30%   (CE endcap subassembly).
13%   (armature subassembly).
56   (including armature
         components).

■ Redesigned features of the new DC motor

The DFA analysis highlighted the problem areas of the old motor design from the handling and assembly standpoint. Through creativity, design for assembly methods and many brainstorming sessions, several ideas matured into the new motor design shown in Figure 4. It was ensured through market surveys that the new design was compatible with customer specifications for material and function as well as process feasibility.

As with the old DC motor analysis, the redesigned motor has been separated into a number of subassemblies to enable a closer comparison between the two designs. The DFA analysis of handling and assembly times was carried out. The overall results are shown in the following summary:

- Total assembly time:
- Overall design efficiency:
- Best design efficiency:
- Worst design efficiency:
- Total number of parts:

171.0 seconds.
26.3%.
37%   (armature subassembly).
15.5%   (housing/flux ring subassembly).
18   (including armature components).

□   **Flux ring subassembly**

The following lists the parts comprising the redesigned flux ring:

(a)   mounting brackets
(b)   flux ring
(c)   magnets.

The gluing process has been changed slightly in the new motor. Rather than using a spring clamp to hold the magnets against the housing during curing, a jig is used. The jig enables the magnets to be easily positioned and clamped to

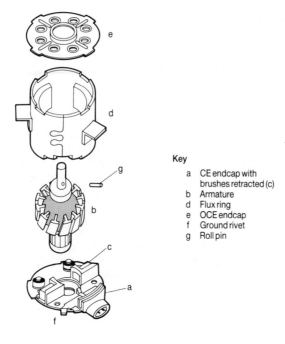

Key

a   CE endcap with
    brushes retracted (c)
b   Armature
d   Flux ring
e   OCE endcap
f   Ground rivet
g   Roll pin

Figure 4   Final assembly of redesigned motor.

the flux ring. In the new motor, the flux ring forms the casing. The mounting brackets are placed in a jig and automatically spot-welded in place. The machine performs all welds simultaneously.

This was the only subassembly in which the design efficiency decreased after the redesign since the number of parts was increased with the addition of the mounting brackets to the subassembly. The method of securing the mounting bracket is more time consuming than the old process; however, it offers more flexibility in dealing with model-to-model variations in mounting bracket design.

☐   **Opposite comm-end endcap**

The spherical bearings were replaced by sealed bearings, to be included as part of the armature subassembly. Therefore, the number of parts in the opposite-comm-end endcap was reduced to a single-stamped metal part. Consequently, this part will be introduced at the final assembly station and can no longer be classified as a subassembly.

☐ **Armature subassembly**

The parts comprising the redesigned armature subassembly are:

(a)   armature
(b)   sealed bearings.

The usage of the sealed bearings has not only eliminated the oil throw washers and washers on the armature subassembly, but also has eliminated the many hard-to-handle components of the bearing in the endcap sub-assemblies. The bearings are symmetric and therefore easy to handle; however, they are difficult to align on the shaft and due to the required interference fit, they cannot be manually pressed on to the shaft. An adequate jig will not only relieve this problem, but will also allow a better control of the bearing-to-bearing dimension in the new design. Additionally, the new bearings are less noisy and last longer. As with the old DC motor, the armature will be considered as pre-assembled for the purpose of the DFA analysis.

☐ **Comm-end endcap subassembly**

Most of the design changes have been made in this area. The parts comprising the comm-end endcap subassembly are:

(a)   comm-end endcap
(b)   brush/pigtail
(c)   spring
(d)   power connector.

The comm-end endcap is an integrated brushboard and endcap made from a special plastic material. The brush tubes are also moulded on to the endcap along with the necessary supports and sockets for the actuating springs and RFI package. The RFI package is not considered for the purpose of comparison to the old DC motor, which was not RFI suppressed.

  The flexible power leads of the old motor have been replaced with a power socket. This socket is separate from the endcap and therefore will provide more flexibility to varying customer requirements while standardizing its assembly to the comm-end endcap. The torsional brush springs will resist nesting and tangling, and therefore reduce handling time. Also, the brush springs double as a brush clip. After insertion of the armature bearing into the comm-end endcap, the brushes are pushed into the commutator to establish necessary contact. This makes the brush actuation a much simpler process and the open design of the comm-end endcap also allows easy access for this operation.

☐   **Final motor assembly**

Each step in the final assembly has been reduced to a simple series of value-added operations (with the exception to the brush actuation). The motor has been designed so that no reorientation is required during the assembly and all components form a stacked assembly. The following parts comprise the final assembly:

(a)   CE endcap subassembly
(b)   armature subassembly
(c)   flux ring subassembly
(d)   OCE endcap
(e)   ground rivet
(f)   roll pin.

The comm-end endcap possesses a deep enough bearing seat that after the bearing (and attached armature) is inserted, no support mechanisms are needed to hold the armature upright for subsequent processes. In the final assembly, the mounting bolts, nuts and washers have been eliminated and replaced with a process whereby the flux ring is crimped to the endcaps. Since a rigid ground terminal is moulded into the connector instead of the flexible leads and terminal clips of the old DC motor design, the terminal is always properly located when the casing is placed on the endcap. The ground connection to the casing is made with a push rivet. The riveting process is an extra operation and the motor is handled in the same manner as for the crimp operation.

■   Discussions

The objectives of the DFA analysis are to redesign the component parts of an assembly such that they are easier to handle and assemble and to minimize the total number of parts in the assembly. After such an analysis is carried out on a product, it may be found that some design proposals are not feasible or that their implementation cost is very high. Beside the redesign proposals, other factors were considered in this motor redesign project such as fabrication feasibility, product development time, development costs, production cost, inventory, customer specifications, etc. Since this motor was designed for the automotive market, a process FMEA, failure modes and effects analysis, was carried out on the designs and assembly processes. This analysis procedure indicated design changes, from the ideal redesigned motor, to decrease the probability of a defective product reaching the customer. In reality, all factors must be considered, in addition to the design for ease of manufacture, and final design compromises may have to be made. In our case, not all trade-offs reduced the new design efficiency. For example, although the sealed bearings were more costly than the spherical bearing assembly, they were easier to handle and assemble and functioned better.

The overall design efficiency of the redesigned motor increased by 7.5%. A 68% reduction in the total number of parts (from 56 to 18) and a 60% reduction in the assembly time (from 424 to 171 seconds) was achieved. The reduction in the number of parts will decrease the in-house inventory and reduce the assembly time. Simplifying assembly operations also decreases the proposed assembly time. The old motor required several procedures to turn the assembly over such as in the crimp of brush tubes and terminal clips to the brushboard and insertion of the bolts at the final assembly. For these operations, extra time was added to the operation and thus design efficiency was decreased. All manual processes in the new design are 'top-down' operations. Top-down assemblies with no hard-to-reach areas and self-locating parts not only benefit manual assembly, but also aid in the planning of automatic assembly stations and the associated tooling (El-Maraghy and Knoll, 1986 and 1987).

Features on the motor which allow for easy alignment of the parts also render this design superior to the old motor. There is a generous chamfer on the plastic CE endcap and a generous radius on the stamped metal OCE endcap to allow ease of alignment for the bearing. The power connector locates on several ribs and settles in a nest in the CE endcap. The chamfers on the upper lip of the endcap allow the flux ring to align itself during assembly to the CE endcap. These features reduce the required handling time for these parts and aid in both manual and robotic assembly.

■ Conclusions

The DFA analysis proved to be beneficial to the overall redesign of the old DC motor. Initial DFA analyses showed where design inefficiencies existed. In these areas, redesign efforts were concentrated on the major design problems and a marked increase in design efficiency resulted.

From the DFA analysis, the design interdependence between the assembly components became clear. This also aids the tooling designers in selecting the appropriate features on which to locate when the fixtures, pallets and fixed tooling are designed. The same thought patterns used in the DFA analysis can also be applied to the design of the assembly equipment to make them simpler and more serviceable.

Together, the reduced number of assembly operations and the smaller inventory of the redesigned motor, a tighter, more efficient assembly line than the old DC model is possible. This will also lower the changeover time and cost between models once batch production is implemented. Cost savings associated with the redesign can increase the manufacturer's share in the fractional-horsepower DC motor market.

References

Boothroyd G. and Dewhurst P. (1983). *Design for Assembly*. Boothroyd and Dewhurst Inc

ElMaraghy H. A. and Knoll L. (1986). *Universal Gripper/Pallet Tooling Design for Motor Assembly*. EHMA-3-C, McMaster University, November 1986

ElMaraghy H. A. and Knoll L. (1987). *Detailed Design of Universal Gripper/Pallet Tooling for Motor Assembly*. EHMA-7, McMaster University, December 1987

Acknowledgements

The motor redesign for assembly is a Co-operative Research & Development, CRD, project, which began in 1986, between Electrohome Ltd and the Centre for Flexible Manufacturing Research and Development at McMaster University. The support provided by the Natural Sciences and Engineering Research Council of Canada, NSERC, to the project through grant number CRD 85-41 is gratefully acknowledged.

3.6
Redesigning mature products for enhanced manufacturing competitiveness

Robert A. Williams
Hewlett-Packard, Loveland, CO, USA

■ Introduction

The trend towards design for manufacture and assembly has advanced swiftly in the past few years, much as the quality movement did in the early 80s. As Henry Stoll (1988) has pointed out, 'design for manufacture represents a new awareness of the importance of design as the first manufacturing step. It recognizes that a company cannot meet quality and cost objectives with isolated design and manufacturing engineering operations. To be competitive in today's market place requires a single engineering effort from concept to production. The essence of the design for manufacture, DFM, approach is, therefore, the integration of product design and process planning into one common activity.'

While there is no question that his statements are true, this author has observed that DFMA, design for manufacture and assembly, today is largely pursued from a new product perspective only, and is seldom applied to the mature or released product. Even on new products, taking giant steps forward in DFMA is often resisted for many reasons. In particular, is the unavoidable pressure to meet the schedule, which leads towards a tendency to choose old and 'safe' approaches. Another reason is the inevitable human characteristic of being resistant to any form of change.

This paper is reprinted by permission of the author from *Proceedings of the 3rd International Conference on Product Design for Manufacture and Assembly*. © 1990 Hewlett-Packard Co.

An alternative strategy to get DFMA rolling is to consider making small, but significant improvements to mature products to learn what works best, then apply this 'new' and 'safe' technique to new products that are coming out of research and development. Whatever the reason, we maintain that ignoring opportunity in an established product arena can be a serious oversight, and can deny a chance to make significant reductions in manufacturing cost, and concurrently, improve quality and profitability metrics.

One of the charters of this paper is to demonstrate that by applying recognized DFMA techniques, a company can improve its competitive position by realizing good returns on the redesign of an established product base. Another purpose is to expose the concepts that had the greatest impacts on our success, while at the same time, reveal some of the pitfalls and obstacles that can erode some of the gains that a cost-reducing team is striving for.

■ Approach and philosophy

One of the first fundamental tenets that we agreed to was that any cost redesign project that had an estimated time to completion of greater than 12 months was automatically eliminated. The basic reasoning behind this was that we felt that we could totally redesign and release most products from the ground up in 18 months, so anything over 12 months was counter-productive to the cost redesign effort.

Our approach initially took on a sweeping investigation of all existing products. For our initial project, we narrowed the field down to one longstanding instrument that we had been shipping for about five years. Since this was a new approach to cost reduction, we wanted to use an instrument that had the most leverage for payback calculations. Our hope was that this would pave the way for any future ones. Once the instrument had been selected, we then proceeded to divide the project possibilities into distinct modular areas of concentration. Each of these areas underwent extensive brainstorming and application of DFMA principles and axioms. While in this process, we looked especially for the following fields of opportunity:

(1) Areas where technology had changed since the advent of the product.

(2) Areas where parts could be reduced by consolidation or by a more clever design.

(3) Areas where there were known reliability concerns that we could concurrently improve with a new design approach.

Once these areas of opportunity were identified, and quantified with respect to their internal rates of return, we prepared a Pareto list and submitted it to functional management for review and subsequent approval.

After the success of the pilot project, the rest of the product base was researched in much the same fashion. The only difference this time was that all the instruments underwent the patterned DFMA scrutiny at the same time, and then a 'total product line' Pareto list was developed to ensure that the highest payback project was attacked first, then on to the next, etc.

■ Execution phase

From the outset of the first project, we adopted and focused on the following principles and philosophy in order to increase our probability of success:

(1)   It was decided that a dedicated engineering staff working on one project at a time was needed to accomplish the goals we had set. We felt strongly that, if engineers' cost redesign efforts were diluted with other projects, our results would be unsatisfactory. One of the keys to success in cost redesign is quick completion. We knew that if we went beyond more than 9 to 12 months, our rate of return would decay rapidly. In fact, we strongly considered that a new design would be better than going through a long redesign. The only time we deviated from the one project focus was when we were waiting for tooling to be completed and prototype parts to arrive for the current project. That time was then used to get the next project off the ground.

(2)   The second execution principle we used was that of avoiding what we call 'creeping featurism'. We had to decide specifically what we were going to do and then execute it rapidly. There is a tendency in engineering circles to keep adding enhancements and features as you go along in a product. This can be upsetting to your schedule and can be reduced by deciding up-front what will be done on the project.

(3)   Third, we had to commit to implement the projects on schedule. We knew that if a major slip occurred, we would jeopardize our 'bread and butter' product lines, creating an undesirable effect on the economics of our division. The two major risks to our schedule were: (a) problems with the new design and (b) production schedule pull-ups. Once you have shut off the purchase of current material and you are inside the required ordering lead time, if you need to increase your production or have to slip your schedule due to a design problem, you will shut down your current production of that product. Also the longer you delay implementation, the more the possible savings are diminished.

(4)   We have to develop a service support strategy early in the project. A

distinction had to be made at the very beginning between 'backward compatible' and 'retrofittable'. The term 'backward compatible' was the one that seemed to cause the most controversy. The service side of marketing usually wants to protect the customer who needs the one small widget that you cleverly redesigned out, in order to restore his vintage product. But at the same time, if you carry all the old widgets in stock for a 10-year service life, just to service that occasional customer, your carrying costs of service stock damage your profitability. Our strategy was to do everything possible to make the new parts backward compatible, but where it did not make economic sense, we chose the retrofittable solution. This is analogous to the carburettor repair kit you buy at an auto parts store when you need a couple of springs or specialty screws to restore your carburettor. While it seems odd, and uneconomical at first, for the parts outlet to sell you extra parts you don't need at the moment, you would be even more perplexed if you had to pay the same three or four dollars for a screw, spring or gasket because of the parts dealer's carrying costs.

(5) Last, we established a product specific requalification plan particularly tailored to the extent of the redesign. We sought agreement, up-front, from our quality and R&D departments, as to what set of environmental requalification tests we should perform. Any failures associated with the redesign would be solved by the redesign project team. Failures that were not associated with the redesign would be reviewed for correlation to product warranty data. If the warranty data supported the qualification failure, then the problem was corrected. If not, no action was taken and the test was assumed to be invalid.

■ Project results

The results of HP Loveland's redesign for cost reduction and enhanced manufacturing competitiveness are broken into five separate products. Some of the products had distinct project divisions themselves. For the sake of simplicity, the results of the redesigns will be listed in consolidated form (Tables 1 and 2).

■ Conclusions – what we learned

Our experiences and conclusions about the nature of redesigning mature products are now beginning to become visible, some four years after starting the first project. Field warranty data has been collected for some time and

Table 1

Redesign project metrics

	Product 1	Product 2	Product 3	Product 4	Product 5
Modular assembly time reduction	44%	28%	18%	38%	38.5%
No. parts reduced	85	61	84	242	135
No. part nos. reduced	69	43	46	130	42
No. operations reduced	200	19	10	53	346

Table 2

Total production cost savings

	Product 1	Product 2	Product 3	Product 4	Product 5
Savings	17.2%	18.6%	7.5%	6.7%	5.3%

then compared with the older designs. We feel very upbeat about the changes that were made, especially since we will probably be building these instruments for at least another 5 to 10 years. The changes have greatly reduced the assembly complexity and the time required to build a unit. This has allowed us to shift some of our 'freed-up' labour capability to newer products that have higher volumes. Parts counts and material costs were improved and the changes have been a plus for our company's profitability and competitive position. General conclusions and redesign tips include the following:

(1)   Do not underestimate the impact of the service strategy on a redesign project. Retrofittability and backward compatibility can be major hurdles.

(2)   Pay particular attention to the total costs of a redesign project. Include costs for any scrap that might be unavoidable, and also factor in things like manual changes.

(3)   Realize that by introducing a new assembly technique or component technology you are running the risk that a new quality problem may appear.

(4)   The commitment must be made early on to stay on schedule. If you slip your schedule, you are certain to be faced with difficult 'no-win' decisions.

(5)   Do everything within your power to prevent your mainstream products from becoming adversely affected in the areas of reliability and availability.

But perhaps the biggest conclusion may be the subjective one that is analogous to how many people felt about our space exploration efforts in the 1960s. While we had an ambitious goal, and finally achieved it, how do you measure the global economic impact that the space race had? If we had not pursued that particular goal, would we still be searching for materials, processes and products that have been on the market now for some time? Only conjecture can offer answers. The fact remains that we did pursue an ambitious and uncharted project, not knowing what all the benefits would be, and we reaped the spin-offs in such an explosive way that no one really can measure it now. In like manner, the biggest benefit of redesigning products for manufacturing competitiveness may not be the bottom line statistics of how much your profitability increased. Rather, it may be the fact that a new culture of 'manufacturing and assembly efficiency thinking' was instituted, without having to take the risks on a brand new project. Putting a dollar value on that benefit is a task even the most accomplished cost accountant would find a challenge.

Reference

Stoll H. (1988). Design for manufacture. *Manufacturing Engineering*, January

Acknowledgements

The author would like to recognize the following personnel for playing key roles in the success of one or more of the described projects: Ric Bacon, Cliff Bergren, Ron Fuhrman, Barb Haas, Steve Hartwell, Robert Hernandez, Bob Hetzel, Bill Jones, Dee Larson, Doug Olsen, Fred Staats and Wayne Willis. Thanks also to Wayne Willis for his able assistance in compiling some of the data in this paper, and for providing both encouragement and editing assistance.

Part 4

Computer–aided DFM Techniques

■ Introduction

Papers included in this part are selected primarily because each one represents an approach at *formalizing* the DFM activity in order to make it suitable for computerization. The result, therefore, of whichever approach is followed, when implemented, is a form of computer-based design system or tool, somewhat similar in concept to an interactive graphics–based CAD or CAD/CAM system, but with an assumed greater degree of 'intelligence' in its modelling scheme in order to represent *manufacturing knowledge* in some way. Indeed, some of the DFM systems described here are designed to link to or integrate with conventional CAD systems.

In terms of general approach, the collective essence or theme of the papers is that they propose a technological solution to the DFM problem. In broad terms, the authors do not propose a managerial or an organizational solution, commonly brought about by getting individuals to work together in multi-disciplinary teams, comprised of design and manufacturing engineers. Nor do they propose an educational approach based on making engineering designers more aware of manufacturing constraints and opportunities. Further, information about manufacturability – principles, rules or knowledge relating product design to manufacturing processes – is not presented as guidelines or checklists, usually of a general nature, to which engineering designers can refer at the appropriate time, as papers in Parts 2 or 3 might propose. Instead, the intention is to encode such design information in the memory of a computer system, so that DFM is part of an overall design activity.

The significance of the term 'computer aided' should not be lost on readers, since this in itself implies a particular approach which distinguishes it from other approaches such as Boothroyd's computer-based methodology (discussed in Part 3) for designing for assembly. With such methodologies, a given product design is examined solely from the assembly aspect of the design – a sort of retrospective look – to see how the proposed design can be modified to ease the assembly process. Some argue that this approach produces solutions that are suboptimal, because they treat design for assembly as an independent problem. In comparison, computer-aided DFM, like the CAD approach in general, implies an integrated approach to solving the design problem, one that considers manufacturing along with other factors (for example, function, quality, cost, and so on) in a more comprehensive and embracing design activity. This means that manufacturability, taken together with the other factors or constraints, is considered at the beginning of the design process, at the conceptual design stage, not when a product design is further downstream. This approach is analogous to simultaneous engineering concepts, currently being developed to reduce the product development time.

Any resulting model of the DFM activity (or of some related issue that is focused on integrating design and manufacture), brought about by some attempt at formalization, is inextricably linked to the means of representation in the computer-based structures that allows the model to be implemented. Thus, for example, a model based on 'if–then' rules (of an expert system) would look different to one that is based on a more rigorous description of parts and related manufacturing processes (of a geometrical model). The effect is that, for papers in this part, it is the means of representation that determine the particular approach taken to produce a solution to the DFM problem, and the various approaches taken in the papers can indeed be categorized by the computer technology used to implement the approach. Of the papers included, three are based on *knowledge-based* (or *expert*) *systems*, one uses conventional *database management systems*, while the other is a progression of CAD/CAM technology with an emphasis on the *data structures* of geometrical models.

An important point to be made is that, unlike papers in other parts of the book, papers here describe development projects and research work, largely taking place at universities. The results of such efforts are prototype systems, whose central aim is to establish the principles and supporting techniques underlying a computer-based system to aid DFM. Their respective degree of completeness in terms of a practical working tool varies from paper to paper. At the time of writing, only one of the papers describes a system currently being offered as a commercial product.

■ Discussion of papers

There are five papers contained in this part. Each presents a different perspective and approach to promoting the practice of DFM, but all rely on some type of computer technology as the enabling technology. The three that adopt the knowledge-based approach refer to their systems as *advisors* or *consultants*. This is consistent with the notion that it is the computer that embodies the expert's manufacturing knowledge, which designers can learn from or refer to. A similar theme is taken by another paper, also designed to act as a consultant, but in this instance more conventional database technology is used instead of knowledge bases. The fifth and last paper concentrates on the geometrical modelling aspects, in what might be regarded as an information-based approach. This paper, although not specifically about DFM, addressing instead computer-integrated manufacturing, focuses on producing an enhanced 'product description system'. As mentioned earlier, for all the papers, the CAD/CAM concept is seen as the natural home in which to place DFM considerations, but only the last paper is specific about extending the CAD/CAM model.

In the first paper, 'A knowledge-based solution to the design for assembly problem', Kroll *et al.* position their approach between the qualitative approach of providing design rules and guidelines (discussed in

Part 2), which are considered too general, and the quantitative approach of assigning numerical codes to parts and assembly operations (discussed in Part 3), which are considered too specific. However, a computer-based advisor, if it is to overcome these basic difficulties associated with 'conventional' approaches, must have to store comprehensive knowledge of the functional structure of parts and of the relevant manufacturing process. A knowledge-based approach makes this feasible, argue the authors, because it is efficient at handling specific *domain knowledge*, can interpret the functional structure of a product and can provide design rules that are tailored to meet specific circumstances.

To represent a product structure in the advisor, the computer model describes the design in terms of features, usually regarded as more familiar to engineering designers, as contrasted with conventional CAD systems which represent parts with solid primitives. This of course poses difficult problems when the intention is to extract features from the geometrical modeller of a CAD system (for example, ambiguity in interpretation, and so on). The paper, however, is based on the premise that such problems will shortly be overcome, and feature descriptions are regarded as the starting point in the use of the advisor system.

In addition to a feature description, the system uses a geometrical and topological description to represent a part. This process is done interactively, with a display module to produce the graphics. From this product structure description, a rule-based 'assembly sequence analyser' generates an assembly sequence by 'exploding' the product structure from 'mating' condition information, contained under 'topology', while applying rules. Practical considerations involving the assembly sequence, difficult to implement in a computer program, are included by checking that certain conditions apply to, say, two parts prior to assembly. This type of information is provided by expert users.

At this stage, the design for assembly activity is entered. Assembly knowledge is contained within the system as five expert modules, each having a knowledge subdomain of the problem. These expert modules, which interrogate the engineer by asking questions and then making suggestions, are arranged hierarchically so that general (company policy level) issues are dealt with first (for example, try to design the product with no subassemblies) before going down, progressively through each module, considering further and further detail. All of this is explained at some length in the paper in the context of a hypothetical electrical alternator, and that example is used to demonstrate the before and after of the design for assembly situation. Although the research conducted by the authors is rooted in axisymmetrical products, they believe that the methodology is more widely applicable.

Further evidence that the advisor-type system is a feasible approach is given in the paper 'Design and manufacturing advisor for turbine disks', by Kim *et al*. The approach identifies the method of representation as important in encoding both design and manufacturing knowledge to ensure that the two are compatible and consistent. Similarly to the Kroll *et al.* paper, geometrical features (of a design) and associated rules are used to link the domains of design and manufacture.

The authors emphasize a design perspective (Kim has produced other work in modelling the design process), and a distinct design methodology to formalize the design to the fabrication process underpins the resulting computer-based advisor. Four phases of design to fabrication are identified: problem specification, feature specification, test and generation. Initially, a design is defined in terms of functional requirements and design constraints (for example, cost, size, performance), after which the design is physically characterized by features – these being generic geometrical entities that are combined to define a given physical design. In the paper, a simplified turbine disk provides the setting for the description, which contains examples of typical features. After a design is specified (functionally and physically), it is evaluated for producibility. The concepts of producible and tractable are defined: a design is 'producible if machines exist which can fabricate the product to specified production constraints'; and a design is 'tractable if it is one that is both feasible (functionally) and producible'. The final phase, generation, is where processes are selected and matched with features to produce a process plan. Sequencing rules (procedural knowledge) comprise the bulk of the expert system knowledge contained in the system and are represented in the form of 'if–then' rules, in common with many other approaches to automate the process-planning function. Knowledge is divided into domain independent (generic knowledge about a larger area of application) and domain dependent (specific knowledge about turbine disks). The prototype system is limited to a simplified structure (containing just five geometrical building blocks) to test the advisor architecture.

Another paper that promotes expert systems is 'Production-orientated design: a knowledge-based approach' by Swift *et al*. It proposes that an expert system, in conjunction with and augmenting a CAD system, is capable of presenting, to an engineering designer, quantitative information (technological and economic) relating to the manufacturability of alternative product configurations.

Similarly to the Kim *et al.* paper, a design methodology underpins the computer-based approach. But in this case it is set in the context of automated assembly, although initially the design methodology has provision to take account of strategic issues by encouraging the designer to consider product rationalization (for example, standardization, identification of families, and so on), which may obviate the need for assembly automation. Such issues are addressed during a functional and a manufacturing analysis. A functional analysis is used to identify components that might' be eliminated (for example, ones that could be combined with others), thus simplifying assembly. A manufacturing analysis is provided to determine manufacturing costs, relating perhaps to changes to a product structure brought about by design rationalization. This latter provision (that is, the ability to anticipate manufacturing costs) is cited as an important feature of the approach towards developing a practical DFM system. Other steps in the methodology relate more to automated assembly and include feeding analysis, gripping analysis and fitting analysis, the latter being used to highlight any fitting processes that are overly expensive. An example of two different stapler designs is used to demonstrate the approach: quantitative data is presented to enable comparison to be made.

The current focus of development is on embedding the knowledge-based DFM evaluator within a CAD system. An implicit assumption here is that manufacturability is largely determined by the geometrical properties of a design, and consequently the majority of the expert rules relate to interpreting geometrical data. Like other approaches aimed at integration into a CAD system, an automated feature-recognition facility is being developed, which in this case would interrogate salient (geometric) properties of a design from an engineering drawing and interpret manufacturability automatically.

A different approach, which also addresses the assembly problem, is taken in the paper 'ASSYST: a consultation system for the integration of product design and assembly system design', by Arpino and Groppetti. Based on comparisons with other DFA methodologies that produce 'suboptimized solutions' for product designs, this paper suggests the need for 'an early interrelated design that leads to a global solution product/assembly system'. Calling this an integrated approach, the authors describe a prototype consultation system, the aim of which is to evaluate a given product design for ease of assembly, including some advice on its redesign, and to provide advice on alternative assembly systems for that particular product design. A user then has the facility interactively to modify these basic assembly configurations and evaluate their respective capabilities.

Unlike the knowledge-based approach used in the Kroll *et al.* paper, this one uses an established database management system to organize the extensive information the system has to structure. Apart from using the more conventional enabling technology, which it can be argued has its penalties, the emphasis in this approach is towards providing alternatives in configuration of assembly system, plus the ability to match an assembly configuration to a product design. Because of the database approach, it is apparently data intensive, also requiring fairly extensive computing resources for support.

In terms of outward appearance, the consultation system is effectively subdivided into a database and a number of application modules (product definition module, assembly definition module, and economic analysis and product/assembly system selection module). Data (product design and assembly systems data) is stored using a hierarchical record stucture, which in the case of product designs records, respectively (at descending levels), product information, subassemblies, components and parts. At the upper-most level of assembly systems data is information on complete assembly systems, such as total number of stations, equipment cost, number of direct and indirect operators, and performance attributes such as volume capacity. Lower levels represent further detail on each alternative system configuration.

In the paper, a case study is presented to demonstrate the workings of the consultation system. A number of figures show the kind of dialogue an engineer would use during an interactive session. An automotive part, a free-wheel, is used as an example. Initially, after individual components are defined and production requirements are specified, the system produces a product structure table. Next, an appropriate assembly system is selected by the user from basic alternative configurations. Any choice is then subsequently evaluated, although this is mainly an economic analysis. An apparent deficiency is the system's lack of graphics or geometrical modelling capabilities, which can restrict its serious application as an interactive design tool and its potential to integrate to a CAD system.

Finally, a CIM (computer-integrated manufacture) oriented approach to achieving design for manufacture, based on the concept of the *product model*, is given in the paper 'Towards integrated design and manufacturing' by Bloor *et al*. The concept is to extend the geometrical model, used as the basis of most CAD/CAM systems, towards a more comprehensive one, referred to as the product model. The additional information largely concerns manufacturing information. The concept of a product model also has another connotation. It is intended to be of *neutral format*, making it easier to transfer product data between different CAD/CAM systems. Such a comprehensive and independent model would support all or most sectors of engineering activity in the life cycle of products.

The communication of information is widely regarded as the core of the CIM approach at integrating design and manufacture, with CIM being essentially a technological, not an organizational, solution to uniting design activity with manufacturing activity. The basis of the approach in the paper is in the *representation* of data, moreover from a database technology point of view. Unlike the previous paper (Arpino and Groppetti), which made use of conventional database technology, the Bloor *et al*. research effort lies in developing an advanced data structure and an interface to allow engineers interactively to define a part, in terms of geometry, material and manufacturing cost.

The paper reports on experience of using an experimental 'product description system', a system to allow product definition data to be created and accessed. The authors' focus is on the information content (data modelling) or the product model, as a means of promoting integration between design and manufacture, rather than on the communications or distribution mechanisms, which are often the centre of attention with CIM development. The essence of the approach is that 'a complete representation of geometry is a basic physical attribute of engineering components, necessary for many applications'. A case study included in the paper demonstrates how integration was achieved for the case of a 2½ D machined component. In common with most approaches in automating process planning, manufacturing methods (cutter data selection, cutter path generation, and so on) were derived by matching features with a database of machining information.

The papers reviewed here reflect the current technological approaches to producing a solution to the DFM problem. All are based on encoding information or knowledge about manufacturability in the internal structures of an interactive computerized design system. The manner in which the information is represented in the computer varies, depending on what approach has been followed. In general, these are known as: knowledge based, database technology or the product model. Of these, there is a widely held view that knowledge-based technology has a number of advantages over its rivals.

As to the effectiveness of a technological approach, there is some evidence that DFM principles can be applied in this way. Of course, to ensure success, use of a DFM consultant or advisor needs managerial support, in order to create the proper climate in which DFM is encouraged; this includes making the designer more aware of issues of manufacturability.

4.1

A knowledge-based solution to the design for assembly problem

E. Kroll,
E. Lenz and
J. R. Wolberg
Department of Mechanical Engineering, Technion –
Israel Institute of Technology, Haifa 32000, Israel

■ Introduction

There are two ways to apply design for assembly: as part of a complete design process where a new product is introduced to meet functional, manufacturing (including assembly), quality and cost targets, or as an independent process, dealing with only the assembly aspect of an existing product. In the latter case, design for assembly is the problem of adapting the product to the proposed assembly process. With the current trends towards flexible manufacturing systems, FMS, it is useful to consider product assembly as early as possible in the design process. The results of such an approach may lead to design changes that could improve the efficiency of the assembly process. The analysis might also lead, for example, to simpler robots with fewer tools and grippers, and less costly fixtures, or indeed, to abandoning robots in favour of alternative process equipment.

There are several categories of products from the assembly viewpoint. For example, many products reveal a layered or stacked composition, where the components are more or less placed one on top of the other. Other products, having base part structures, often consist of such stacked sub-assemblies mounted on the relatively large base- or frame-component. We

This paper is reprinted by permission of The American Society of Mechanical Engineers from *Manufacturing Review*, Vol. I, No. 2, June 1988, pp. 104–8 have chosen in this paper to concentrate on those products that exhibit axial symmetry, for example, like electric motors.

It turns out that the parts and subassembly composition of a product has considerable effect on its ease of assembly, even for apparently simple cases. Trivial operations for people, like turning a group of parts upside down, might prove extremely difficult, if indeed even possible, for some robots to perform. Often, such operations, when performed manually, are executed incorrectly, leading to assembly errors.

■ Conventional versus knowledge-based approaches

Two methods for guiding product design have been suggested and implemented in the past: a qualitative approach which presents the designer with general rules and guidelines accompanied by illustrated examples (Andreasen *et al.*, 1983), and a quantitative one which assigns time periods, costs and numerical codes to various part characteristics and assembly operations (Boothroyd and Dewhurst, 1983). The first is often considered too general to be practically applied during design. Merely presenting the guidelines to the engineer, whether on paper or on a computer monitor, falls short of providing a useful methodology. Moreover, general rules are by nature more prone to misinterpretation, and the lack of any comparison criteria makes evaluation of alternative designs difficult. The quantitative method requires very specific information such as the expected production rate, the cost of assembly hardware and symmetry properties of components. It involves quite a tedious and time-consuming process of completing standard worksheets based on standard coding charts and formulae. This approach lacks flexibility by relying on 'reference' data which might not be available to the designer at the time of analysis, and in using 'standard' assembly equipment as a basis for comparison. But its two main drawbacks are the implicit way of identifying design improvements and, even more fundamental, its inability to treat products at a higher level than the individual parts. As a result, configurative design can only take place by elimination or integration of parts.

Automatically advising the designer about how to improve the product must be based on a comprehensive knowledge of two areas: understanding the functional role of each part, and having an adequate model of the relevant assembly process with regard to operations, tools, and so on. But acquisition of such information by a computer program poses some difficulties. First, the amount of knowledge is enormous. Second, it consists of formal as well as heuristic aspects. And thirdly, both the domain knowledge and the particular circumstances of the design process may be constantly changing: designing a product to be assembled by a simple

SCARA-type robot, for example, could not be the same as for an assembly cell with a pair of six degree-of-freedom robots. Moreover, such a program should also simulate decision-making stages of human design processes. But design is certainly not a well-understood process, being hard to formalize and represent, and difficult to solve systematically (Simon, 1981).

The rapid development during the past few years of artificial intelligence techniques, especially 'expert systems' or 'knowledge-based systems', provides us with powerful tools for handling such tasks. By efficiently utilizing specific domain knowledge, a computerized consultant system can 'understand' high-level issues concerning the structure and character of the product, and the goals of the specific design process, and provide design guidance accordingly. A functional model of the product can be created, serving as the basis for devising new, easier-to-assemble configurations. Elimination and integration of parts can begin when standard and unique components have been identified, while individual part design is facilitated by recognizing the geometric features of the system. Moreover, the explanation and justification facilities incorporated in many knowledge-based systems, together with the separation of the reasoning mechanism from the knowledge base, are extremely useful for an advisory system. Before describing the system which is the subject of this paper, the question of how to represent the product throughout the design process is addressed next.

◼ Product representation

Although common practice for constructing the main body of knowledge in an expert system is to use rule-based programming, two other issues concerning knowledge representation remain to be solved: What should be the starting point for design, and what data structures will be used to describe the product throughout the process?

Ideally, the description of a product is stored in a computer-aided design, CAD, database. But contemporary CAD systems usually do that at a much lower level than what might be considered convenient for design tasks and human communication. For example, engineers sometimes do not think in terms of solid volumetric primitives, but rather they may specify features having engineering meaning like 'equally spaced holes on a pitch circle', 'V-belt pulley', and so on. One obvious way to bridge this gap is to extract feature information from the CAD data, although this has been shown to be a very difficult process with occasionally ambiguous results (Henderson, 1984). On the other hand, the newer generation CAD systems are supposed to support the required high-level interaction (Brody, 1987). Consequently, in this study we decided to start with natural feature-based engineering descriptions for the parts, which the user creates with the help of a pre-prepared 'library' of features.

Besides defining the exact geometry of each part, other information is required. First, a topological characterization of the product is entered by the user in terms of several 'standard' mating conditions (for example, 'above', 'tightly-fits-in') to define relationships among parts. Then, administrative data such as the material, and whether the part is a standard component, are obtained from the user.

Having a thorough description of the product does not mean such information is needed at all times. The human design process is characterized by having varying levels of product abstraction at different stages. Starting with very general concepts of the product, the description is gradually refined by including more and more details in it. As described in the next section, the present system can accommodate various representations.

■ System architecture

The system software consists of nine programs or modules, two library files, two working memory files, a connection to another computer and a main program to control the whole design process. The program has been implemented in Turbo-PROLOG on an IBM-XT. The file structure is schematically shown in Figure 1.

Normally, an existing or a new design description is entered

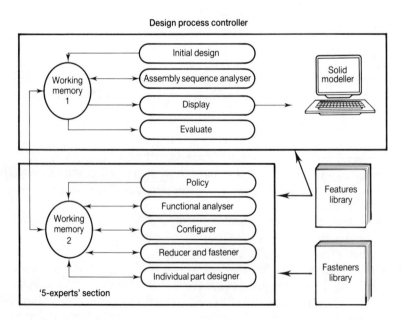

Figure 1 Architecture of the design for assembly system.

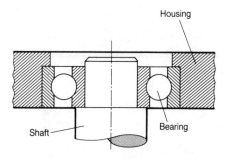

Figure 2   A ball bearing, housing and shaft.

interactively using the initial design module. Each part constitutes a 'frame' with 'slots' for the following attributes: serial number, name, quantity, type (for example, standard screw, standard rolling bearing), material, geometry and topology. Some of the 'slots' are filled with the aid of the features library file. This file contains definitions of 'basic bodies' (for example, hollow cylinder, cup, hex nut) and shape features like eccentric-threaded-hole, cylindrical-cavity, and so on. When choosing a certain feature from the menu, the user is prompted for the required parameters of that feature, with default values shown as well. For example, the frame-like data structure for the bearing in Figure 2 might be:

FRAME TYPE:	part
NUMBER:	7
NAME:	6304 ball bearing
QUANTITY:	1
TYPE:	standard rolling bearing
MATERIAL:	irrelevant
GEOMETRY:	basic body: rolling bearing (50, 20, 15)
	features: 'outer ring', ring (52, 40, 15, 0)
	'rolling elements', ring (40, 30, 15, 0)
	'inner ring', ring (30, 20, 15, 0)
TOPOLOGY:	above ['inner ring', shaft ('shoulder')]
	tightly-fits-in ['outer ring', housing ('bearing mount')]
	assembled location: world (0, 0, −37.5)

Notice the material slot was filled with 'irrelevant', as the part is of standard type and we do not wish the system to suggest combining it with the housing or the shaft, an advice that required identical materials. The geometric description consists of the basic body from the library, with suitable parameters (OD, ID and width), and features with corresponding values. Note the definition of the rolling elements feature as a 'ring', which is actually the shape of the envelope, but is satisfactory for our purposes. The topology slot defines the mating conditions of each feature with other part features. What

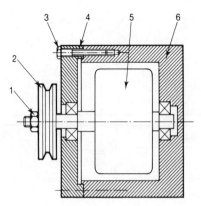

Figure 3 A hypothetical product with potential assembly problem.

seems to be missing, like the outer ring being 'below' the housing is actually entered as an 'above' condition when the housing is described. Finally, the location of the bottom centre point of the part's basic body is given in 'world' Cartesian co-ordinates, arbitrarily set for the assembled product.

The next step is to analyse the product from the assembly sequence viewpoint. This is accomplished with the aid of the rule-based assembly sequence analyser module. First, the exploded view of the product is created from its assembled description now located in the working memory file. The program uses the mating conditions and other information to sort the parts and applies rules such as: A bolt with its head up will be 'exploded' higher than the part located below the head. A series of such rules leads to an ordered list of part names representing the exploded view of the product, which is shown by the display program. Contrary to what one might expect, the exploded view is not the solution to the assembly sequence problem, as individual parts may be combined to form a subassembly before joining into a final assembly. The second stage of the analysis commences as follows: from the mating conditions in the description, all pairs of parts that are fastened together are identified. Sublists of the exploded view list between each such pair constitute potential subassemblies because they represent groups of parts held together by the first and last parts of the list.

But not all potential subassemblies created in this way satisfy practical constraints, as demonstrated by the hypothetical product in Figure 3. There, parts 3, 4, 5 and 6 could theoretically be considered a subassembly, but if we then try to assemble parts 2 and 1, it would be difficult to hold part 5 against the tightening torque applied to the nut, as part 5 is now enclosed by parts 4 and 6. Such understanding is usually referred to as 'common-sense reasoning', widely available to humans, but hard to implement in a computer program. Our analysis approach uses such capabilities to screen out some of the potential subassemblies, and proceeds with identifying the direction from which the others are to be assembled. This involves checking two conditions: a suitable part must exist at either end of a subassembly to serve as a base part which goes first to the assembly fixture, and all other parts in that subassembly must be 'stacked', or have 'above' and 'below' mating conditions with their neighbours. The whole process is recursive, so a possible assembly sequence generated by the system will ultimately consist of an ordered list; some of its elements are individual part names, and others might be ordered lists themselves, representing subassemblies, which are also composed of parts and subassemblies. Besides the assembly sequences, other information extracted so far (like identification of the fasteners in the product, which subassemblies are possible and which are not, and the assembly direction for each subassembly) is also added to the working memory.

As one can hardly imagine an engineering design process carried out in words alone, provision is made through the display module to draw the product on a graphics terminal. Each part's basic body is converted there to a constructive solid geometry, CSG, description. Then proper scaling is performed, sometimes forcing the exploded view to be drawn in two or more 'stacks', and the whole CSG file is sent via a communication link to another computer running a conventional solid modeller.

At this stage the user makes a decision whether actually to start the design for assembly session with the system. If so, the evaluate module is bypassed, the current working memory is saved to serve as the basis for improved designs, and the system activates its '5-experts' section arranged in a hierarchical 'blackboard' architecture. This means that several programs, simulating 'experts' in subdomains of the problem, share a common working memory, or 'blackboard', where the design evolves. These 'experts' have a distinct hierarchical ranking among them, with each module treating the product at a lower abstraction level (or higher detail level) than its predecessor.

The first of these programs, the policy module, interrogates the user as to the general goals of the design process. This stage simulates company policy level decision making, and produces goals to be satisfied later. Such policy goals might be: 'try to design the product to have no subassemblies' or 'do not change part X as it is an interfacing part to something else'. Besides these general goals, the program also asks the user whether the product could be made a 'throw-away' type of product, or should it be easily dismantled for maintenance and repair. This has a significant influence when choosing fastening methods. Before the design can commence, functional analysis (as opposed to the structural analysis performed earlier) should be carried out. This is accomplished by the functional analyser module, which employs extensive knowledge about different machine elements and their functions, in order to distinguish between parts that perform supporting functions such as fasteners and spacers, and functionally important parts (for example, housings, bearings, shafts).

The third 'expert' module, the configurer, plays perhaps the most important role in the design process: making the initial decisions regarding the structure of the overall product. As with human design processes, the earlier the decision, the more significant is the effect that it will have on the results. Based on the functional model of the product created earlier, the program attempts to plan new configurations as instructed by the policy level goals. This involves a complex procedure of checking the structure of the product to pinpoint conflicts between it and the intended simpler assembly sequence, presenting these findings to the user and suggesting remedies, kinematically sorting the functional model of the product (for example, a stationary housing and a rotating shaft should be connected via a bearing), hypothesizing about 'stacking' and 'fastening' rquirements, checking the latter against heuristic-constraining rules (for example, a rolling bearing cannot be simultaneously assembled on the shaft and housing) and, finally, establishing goals to be satisfied by consecutive modules.

Down a level in the hierarchy, the reducer and fastener 'expert' is responsible for two essential functions in any design for assembly process: reducing the number of parts in the product and choosing suitable fastening methods. Although no truly superfluous components should exist in the product, some parts (for example, spacers) might be eliminated if their function can be incorporated as features of other parts. The program now uses information accumulated during previous stages, together with its own knowledge base, to make sensible recommendations such as integrating bearing-support spacers into shafts as steps, or into other parts, but not into the bearings themselves. The second task, choosing the fasteners, is somewhat more difficult. Here, previous goals define both the fastening needs and the product character (for example, easily disassembled, throw-away). The program consults a library file of fastening methods with their characteristics, picks the potential ones, sorts them according to their ease of assembly and asks the user to indicate a preference. When the choice is made, that fastener is incorporated into the product model and goals are accordingly established for the next module.

The last 'expert', the individual part designer, embodies all previously accepted recommendations in each part design. Although some rules can accomplish this automatically, the program is highly interactive, consulting the user at each step. To some extent, this allows the human designer's creativity to participate in the process, after the intention of the program became clear. Besides satisfying the already established goals, general improvements in the handling, orientation and assembly characteristics of each part can take place here. These improvements include turning asymmetric parts into symmetric ones, or exaggerating their asymmetric features, adding chamfers and tapers, 'closing' the ends of springs, thus preventing parts from tangling and nesting during feeding, and so on.

The product evolution by the last three 'expert' modules includes the capability to backtrack. When an 'expert' module cannot satisfy goals dictated to it from further up in the hierarchy, the cause of failure is returned via the main control program to the higher 'expert' for reconsideration of those decisions. If all goes well, the design process controller activates the assembly sequence analyser and display modules. Then the last program, the evaluate module, compares the original product design with the improved one, and produces a summary of the assembly needs such as the required operations, fixtures, tools and grippers, and the sequence with which the product is to be assembled.

■ Structural design example

The following example demonstrates structural changes made to a product to facilitate easy assembly by a single robot. Clearly, the tools described in this paper do not presuppose robotic assembly, which is used in this example only for purposes of illustration. Figure 4 is the assembly drawing of an electric alternator whose description has first been entered to the system. An assembly sequence analysis of the product showed that no 'simple' order (with no subassemblies) was possible, and then suggested the following operations:

(1) Mount the 'front housing' (part 6) on a fixture, add to it the 'front bearing' (7), the 'retainer' (8), and then tighten the three 'short screws' (9) to form the first subassembly.

(2) Place the 'rear housing' (13) with its open side up on another fixture and insert the 'rear bearing' (12) in it, to create a second subassembly.

(3) To assemble the product, start with the 'nut' (1) in a fixture, add parts 2, 3, 4 and 5, then assemble the first subassembly (parts 6, 7, 8 and 9), then parts 10 and 11, turn the second subassembly (parts 12 and 13) over and place it, and finally tighten the 'through bolts' (14).

In the design for assembly mode, the user indicated to the policy module the desire to assemble the product with the simplest robot possible, and maintain the easily disassembled character of the product. The former led to establishing the goal: 'try a "simple" assembly sequence', which means avoiding subassemblies. Then the functional analyser module was called upon, to identify parts 3, 4, 6, 7, 11, 12 and 13 as 'functional parts'. This involved the establishment of goals such as: 'the "rotor" (11) is a shaft mounted between two bearings, parts 7 and 12'. It also involved the creation of findings such as: 'the "front bearing" (7) provides bidirectional location of the "rotor", and the "rear bearing" is a non-locating bearing'. The next program, the configurer, had trouble trying a 'simple' assembly sequence, so its first suggestion to the user was to change to a cross-location arrangement

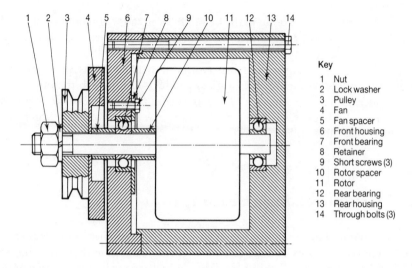

Figure 4 The original alternator: two subassemblies are required for an assembly sequence that starts with the 'nut'.

of the 'rotor' in the bearings (that is, each bearing locates the shaft in another direction). The user, judging that such an arrangement is acceptable for a relatively short shaft with low thermal expansion, responded affirmatively. With that, a 'simple', 'upside-down', assembly sequence was found to be possible if a step was to be incorporated in the 'rotor' shaft to support the 'rear bearing', along with the two spacers (parts 5 and 10) added to the product model. In attempting to reduce the number of parts in the product, only the two spacers proved to be candidates for elimination, from which the user found it difficult to integrate the 'fan spacer' (5) in another part, but accepted the recommendation to eliminate the 'rotor space' (10) and introduced a step in the shaft in its place. Then, the system suggested suitable fasteners for an easily disassembled product, from which the user chose screws for the housings, and splines with a retaining ring for the pulley and fan. A drawing of the improved alternator which permits a 'simple' stacked assembly, starting with the 'rear housing' (part 13), is shown in Figure 5. The next obvious stage of treating the handling and orientation of each part was omitted from the current example because of the similarity to conventional design for assembly techniques.

Comparison of the original design with the new one shows that the parts count was reduced from 18 to 12 (the latter could have been even less, had the user chosen to replace the three bolts with another fastening method). Furthermore, the three fixtures that were needed, one of them quite complicated, were replaced by one simple fixture, and the relatively difficult 'flip–over' operation was eliminated, with all remaining assembly operations

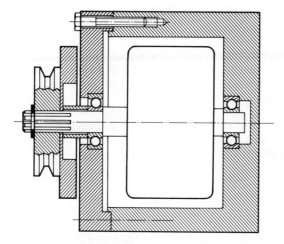

Figure 5 The improved alternator: no subassemblies are required for an assembly sequence that starts with the 'rear housing'.

executable by a four degree-of-freedom robot. From the part manufacturing viewpoint, only the 'rotor' is now more complicated to produce, having additional steps, splines and a retaining ring groove.

■ Conclusion

The emerging technology of 'knowledge-based systems' enables an unconventional type of design aid, where suggestions to the designer are tailored to suit specific circumstances and product character. An application to the design for assembly problem has been described, suggesting the potential of such an approach to solving other conceptual or configurative level engineering problems. To have a truly 'expert' system, the essence of the knowledge base needed for the task and the method of representing this knowledge should be thoroughly studied before implementation, thus ensuring a general-purpose structure with incremental expansion of the knowledge base. Although the current system has been limited to the axisymmetrical family of products, the methodology introduced is believed to be generally applicable. Within the current framework, additional geometric features and fastening methods can be added to the library files, and the rule-based knowledge bases could be enhanced with more specific rules. Moreover, having devised an easier-to-assemble configuration for the product, design can proceed with more conventional design for assembly techniques like those of Boothroyd and Dewhurst (1983), which emphasize individual part treatment.

References

Andreasen M. M., Kahler S. and Lund T. (1983). *Design for Assembly*. Berlin, Heidelberg, New York, Tokyo: IFS (Publications) Ltd and Springer-Verlag

Boothroyd G. and Dewhurst P. (1983). *Design for Assembly – A Designer's Handbook*, Dept. of Mech. Engr., Univ. of Massachusetts, Amherst

Brody H. (1987) CAD meets CAM. *High Technology*, **7**(5), 12–18

Henderson M. R. (1984). Extraction of feature information from three dimensional CAD data. *PhD Thesis*, Purdue University

Simon H. A. (1981). *The Sciences of the Artificial* 2nd edn, Cambridge, MA: MIT Press

4.2

Design and manufacturing advisor for turbine disks

Steven H. Kim, Stephen Hom and Sanjay Parthasarathy

Knowledge Systems Program, MIT Laboratory for Manufacturing and Productivity, Cambridge, MA 02139, USA

■ Introduction

A good design must incorporate manufacturing knowledge to be cost effective and reliable. Since designers are not manufacturing experts, however, they often create designs which cannot be easily fabricated. The design and manufacturing advisor, DMA, is a generic architecture for an expert system which aids designers by integrating design and manufacturing processes (Kim, 1987).

DMA fulfils its role by incorporating knowledge of the relevant requirements and the limitations of all production stages from fabrication to assembly and inspection. It provides the designer with production knowledge throughout the design process and thereby helps to design the final product. A *feasible* design is one that satisfies its functional requirements and constraints, while a *tractable* design is one that is both feasible and easily produced.

This paper introduces the concept of an overall design procedure, a generic process which uses manufacturing knowledge to produce high-quality designs made to specifications. This overall design procedure represents an integrated design-to-fabrication process.

This paper is reprinted by permission of Pergamon Press from *Robotics & Computer-Integrated Manufacturing*, Vol. 4, No. 3/4, pp. 585–92

To demonstrate the feasibility of the DMA, we have implemented its basic approach in a specialized package called the turbine disk advisor, TDA. This is a system to assist in the design of turbine disks for aircraft engines.

The TDA incorporates a design procedure which ensures the satisfaction of specifications at every step. The system was developed for the following reasons:

(1) to identify the methods by which expert manufacturing knowledge can be used to improve design;

(2) to determine the implications of integrated design and manufacturing;

(3) to improve co-ordination between design and manufacturing departments.

One major hurdle in the software development effort lay in finding a method of representing both design and manufacturing knowledge so that they are compatible and consistent. This paper describes the use of features and rules to span both realms. Features are generic geometries which constitute the building blocks from which a final product can be developed, while rules depict relationships among various objects. The TDA facilitates further expansion to incorporate additional expertise as well as new software techniques.

■ Steps from design to fabrication

The TDA incorporates four distinct phases in the overall design procedure, from the conception of the design to the generation of a list of operations (Figure 1).

These phases are:

(1) Problem specification.
(2) Feature specification.
(3) Test.
(4) Generation.

The following subsections describe each phase in greater detail.

□ Problem specification phase

The problem specification phase, which defines the design problem, is the first step in the overall design procedure. Problem definition represents a critical aspect of the process; it must be thorough without being superfluous.

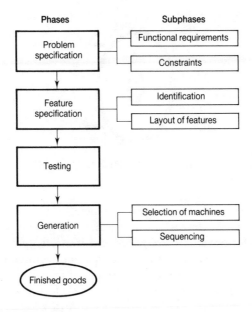

Figure 1   Phases from design to fabrication.

This phase has implications for the entire process, as the system seeks to meet specifications at every step.

This phase consists of identifying the functional requirements, FRs, and constraints of a design. FRs represent the minimum set of independent specifications which completely defines a problem (Suh, 1984). Once the FRs have been enumerated, the next step lies in determining design constraints. A design is often limited by cost, size and performance considerations. These constraints reduce the number of feasible designs and thus help direct the design process.

☐ **Feature specification phase**

A physical design can be characterized by a combination of features. Figure 2 shows how a simplified turbine disk, for example, may be represented by its features.

The concept of features can be defined as follows: *features are generic geometries which can be combined to define a physical design.* This definition rests on the idea that any object can be built up from standard geometries. Examples of such geometries in three dimensions are: cube, cylinder, sphere and cone; and in two dimensions: square, rectangle, circle and triangle.

Once the functional requirements of a problem have been specified, these must be satisfied by an actual design. This activity is called the feature

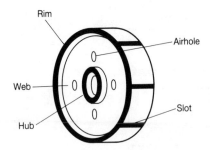

Figure 2   Simplified turbine disk.

specification phase, which consists of the identification and layout of physical features. Satisfactory features are identified according to the relationship between each feature and the requirements it meets.

Consider two cases of turbine disks: one required to operate at high temperatures (disk A) and the other required to operate at high pressures (disk B). Both these disks possess certain basic features (such as rim, web, hub and slots). However, the high-temperature disk (disk A) will have a web of greater surface area than disk B. In addition, disk A may have airholes whereas disk B may not. Thus there exists a correspondence between requirements and features.

The specified features must be arranged so that the product can fulfil its function. The designer determines the relative positions of the features, thereby constructing the final product from basic geometries.

☐   **Test phase**

Once the design has been specified by the locations of the features, it is ready for evaluation. A design is deemed *tractable* if it represents the intersection of three spaces: specifications, features and producibility (Figure 3).

Thermal and stress analyses are used to ensure that the design conforms to functional requirements and constraints. A design is tractable only if it is constructed from features appropriate to the domain of turbine disks (for example, a disk designed with a square hole at the centre will not pass the evaluation stage). As shown in Figure 4, each feature is examined to ensure its relevance to the design domain.

Testing for producibility requires information on production constraints associated with the final product. A design is producible if machines exist which can fabricate the product to specified production constraints. Thus the design is tractable if it fulfils specified requirements, can be produced and is built up from features appropriate to the design domain.

The first three phases of the overall design procedure correspond to the

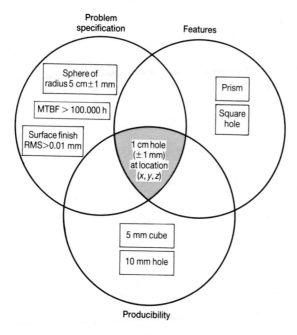

Figure 3   Intersection of three spaces.

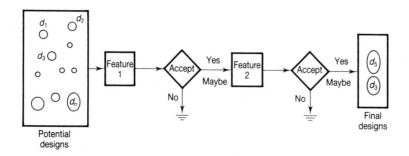

Figure 4   Evaluation of features.

three spaces in Figure 3 as follows. The space of specifications can be seen as the definition of the problem; obviously, an appropriate design must satisfy the defined problem. The space of features concerns the representation of the problem; the design must be represented correctly. The space of producibility describes the ability of a manufacturing system to fabricate the design.

□   **Generation phase**

Once a design is deemed tractable, it enters the generation phase for developing a production plan. This plan defines the operations necessary for product fabrication as well as their relative sequence. Decision rules for selecting operations attempt to match design requirements to machine or process capabilities. Operations selection is thus closely linked to feature representation.

Rules for sequencing operations are developed using various tools such as the fishbone technique (Ishikawa, 1976), which organizes knowledge into cause-and-effect diagrams; sequencing rules comprise the bulk of the expert procedural knowledge in the system. Procedural knowledge is presented in the form of if-then rules, as illustrated in Appendix A.

■ System expertise

DMA accommodates the full spectrum of knowledge in areas ranging from design to fabrication. Figure 5 shows a generalized network architecture supporting DMA: a specific application would incorporate relevant knowledge and databases. Ideally, a system should encode all information which might affect the design process. The challenge lies in identifying the most appropriate information, given resource limitations.

The knowledge required for a DMA is of two types: domain dependent and domain independent. Domain-independent knowledge, as the name indicates, is generic knowledge which applies to more than one specific area within a larger field of application. For example, rules which specify appropriate uses for finite element analysis are domain independent within the larger realm of mechanical design. Domain-dependent knowledge is more specific and consists of knowledge pertaining to a given domain (such as that of turbine disk design).

The advantage of classifying knowledge into dependent and independent types is that it enhances modularity to provide a more flexible system. Thus a system currently used to design turbine *disks* can be used to design turbine *blades*, simply by exchanging a small subset of the domain-dependent knowledge modules.

Knowledge can also be classified as either *declarative* or *procedural*. Declarative knowledge consists of feature and machine representations and of facts about materials, machines, functional requirements and constraints. Procedural knowledge encompasses rules concerning process selection and sequencing. In addition to rules and facts, the system also includes analyses (such as thermal and stress analyses) which are used in the test phase.

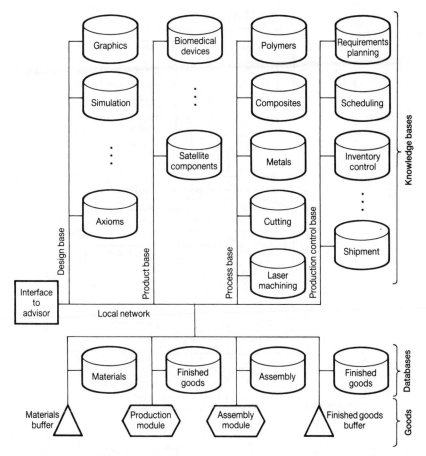

Figure 5   Network configuration for the design and manufacturing advisor.

■ Software architecture

The network configuration for DMA (Kim, 1987) is given in Figure 5, while its system architecture is presented in Figure 6. The system contains two main modules: the domain-independent module, DIM, and the domain-dependent module, DDM.

The DIM can be decomposed into four parts:

(1)   Primary level.

(2)   Secondary level.

(3)   User interface.

(4)   Error handler.

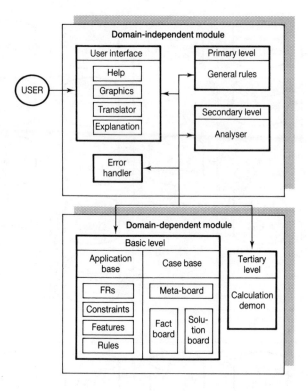

Figure 6   Software architecture.

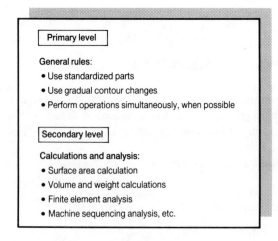

Figure 7   Domain-independent module.

Tertiary level

- Serves as a gateway to domain-independent module
- Calculation demon:

  Defined as an agent triggered by an external stimulus

  Usually a procedure activated by a detecting component

Basic level

Application base:

- Functional requirements, constraints and features applicable to the domain
- *Sample rules:*

  Heat treat after machining

  Inspect after every cutting operation

Case board:

- *Fact board* contains the functional requirements, constraints and features applicable to the particular product being designed
- *Solution board* keeps track of the evolving design
- *Meta-board* provides information such as level of explanation appropriate to a particular case

Figure 8 Domain-dependent module.

The primary level contains both declarative and procedural knowledge, while the secondary level contains procedures for analysing the design. Figure 7 illustrates the types of knowledge found in DIM.

The user interface provides for friendly, intelligent interactions with the user. The error handler addresses exceptional cases, such as users' attempts to drive the system beyond its intended capabilities.

DDM contains knowledge applicable to the domain of turbine disks; it can be divided further into the basic and tertiary levels, as shown in Figure 8. Within the basic level, the application base contains procedural and declarative knowledge about disks, while the case base keeps track of the disk under design. The application base also stores a set of icons representing the basic features of disks.

Knowledge representation

The turbine disk advisor's basic representation paradigm consists of frames (or templates), which unify procedural and declarative expressions of knowledge. A *frame* is a structure consisting of a list of slots and entries. Each

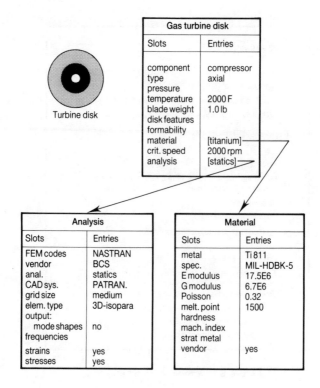

Figure 9   Frame of simplified turbine disk.

entry contains a specific value or a procedure for obtaining the value. A value in a slot can be manipulated, and the result propagated throughout the logical structure of the knowledge base. Alternatively, a slot may contain a procedural specification – that is, a set of instructions to compute a value for the slot. These instructions may combine information from other slots and frames.

Figure 9 shows a frame for a simplified turbine disk. Some slots specify properties associated with the turbine disk, such as the notation that the disk is a component of the compressor stage. The entries for certain other slots may include default values, such as the fact that the disk is part of an axial–flow configuration.

Still other slots may refer to external frames. For example, the analysis slot of the gas turbine disk frame in the figure refers to an analysis frame to compute stresses. The former frame also contains a 'material' slot whose entry is a pointer to a material frame. The material frame holds information on the properties of a specific metallic alloy, such as its modulus, hardness, strength and melting point.

■ Implementation of the turbine disk advisor

As described previously, the generic DMA involves four steps from design to fabrication. These four steps are simplified in the current implementation of the TDA.

During the problem specification stage, the system recognizes two sets of independent functional requirements: pressure and operating speed, or temperature and operating speed. Since design possibilities are limited even from the start, the space of tractable designs has not been restricted by further constraints.

The feature specification phase is simplified by restricting the features to five basic building blocks, namely: rim, web, hub, airholes and slots. Each of the two sets of functional requirements (pressure and speed, or temperature and speed) pertains to three templates (or collections of basic features).

Each of the three templates consists of a rim, web and hub, plus a variable number of airholes and slots. The user may specify the positions of the airholes and the relative positions of other features.

During the test stage, the system checks whether the design fits the intersection of the three spaces pertaining to tractability. A design satisfies the specifications if it passes both stress and thermal analyses. The features constraint is satisfied if the disk incorporates the basic features relevant to the design domain. The design lies in the space of producibility if the final product can be produced to specifications; this condition is tested by checking frames for features with those for production machines; a match indicates a successful design.

During the generation stage, operations are first selected by matching associated frames for features and machines. Then operations are sequenced by activating a series of decision rules, many of which are implicit in the structure of the program itself.

Appendix B gives a sample interaction for the turbine disk advisor. At the end of the interaction, successful designs are displayed along with the corresponding sequence of operations.

■ Discussion

☐ Limitations of the current system

The prototype turbine disk advisor incorporates a simple structure designed to demonstrate the feasibility of the DMA architecture. The capability of the system has been limited by restricting the space of features and the number of feasible designs. In general, the simplest rules and procedures have been implemented.

In a more practical system, some elements would require elaboration (for example, the module for stress analysis would need to be replaced by more sophisticated finite element analyses). However, structural modularity facilitates the replacement of simple routines with more involved ones.

A graphics interface based on iconic feature representation is currently being developed to improve the ease and flexibility of user interface. Icons representing features appropriate to the application domain are associated with the application base. These icons can be manipulated in conjunction with dialogues and menus while using the keyboard and mouse as input devices.

☐ Future work

Several areas related to design and manufacturing integration deserve further attention. Over the next few years, these areas include learning capability, intelligent knowledge acquisition, multi-level explanation and natural language interfaces. Over the long term, research should address the representation of space and time, generic problem-solving techniques for design and advanced graphics capabilities to provide for a 'language of pictures'.

☐ Conclusion

The architecture of the design and manufacturing advisor provides a framework to encapsulate different types of knowledge needed for the integration of the design and production functions. A key advantage of the architecture lies in its modularity, which allows for flexibility and rapid prototyping by exchanging software modules.

The utility of the DMA as a generic framework has been validated through the implementation of a simple package called the turbine disk advisor. The full capabilities of an integrated design and manufacturing advisor, however, will depend on continuing efforts in the development of conceptual and software tools.

References

Boyce M. (1982). *Gas Turbine Engineering Handbook*. Texas: Gulf

Clocksin W. F. and Mellish C. S. (1984). *Programming in Prolog*, 2nd edn. New York: Springer

Harris J. A., Sims D. L. and Annis, C. G. Concept definition: retirement for cause of F100 rotor components. AFWAL-TR-80-4118

Ishikawa K. (1976). *Guide to Quality Control*. Tokyo: Asian Productivity Organization

Kim S. H. (1985). *PhD thesis*, MIT May

Kim S. H. (1987). A unified architecture for design and manufacturing integration. Technical Report, Laboratory for Manufacturing and Productivity, Cambridge, MA: MIT

Kim S. H. and Suh N. P. (1985a). Application of symbolic logic to the design axioms. *Robotics Computer-Integrated Mfg*, **2**, 55–64

Kim S. H. and Suh N. P. (1985b). On an expert system for design and manufacturing. *Proc. COMPINT '85*, ACM and IEEE/Computer Society, Montreal, Canada, Sept

Meethan G. W. (ed.) (1981). *Development of Gas Turbine Materials*. New York: Halsted Press

Suh N. P. (1984). Development of the science base for the manufacturing field through the axiomatic approach. *Robotics Computer-Integrated Mfg*, **1**

Acknowledgement

We wish to acknowledge the help of our undergraduate assistant, Fred Chong, for his contributions to the development of the turbine disk advisor. Fred played a vital role in developing the PROLOG implementation of TDA.

■ Appendix A: sample rules

The following serves as an illustration of the type of decision rules used in the turbine disk advisor.

☐ Space of features

/* The first rule states, for example, that the rim is acceptable if its inner diameter equals the web's outer diameter and if the width of the rim does not exceed the maximum width of the disk. Implementation of the space of features has been simplified, as explained in the body of this paper, so that all included features are tractable. */

- a rim is acceptable if (1) inner diameter of the rim = outer diameter of the web
  and   (2) width of the rim ≤ maximum width of the disk.
- a hub is acceptable if (1) outer diameter of the hub = inner diameter of the web
  and   (2) width of the hub ≤ maximum width of the disk.

☐ **Space of producibility**

/* The design is producible if selected features can be fabricated to specifications. Testing for producibility is achieved by matching the feature frame with the machine frames. A match indicates a successful design. */

- Airholes are producible if hole diameter ≥ smallest of diameters available.

☐ **Space of FRs and constraints**

/* Thermal and stress analyses check whether the design satisfies the requirements and constraints. */

☐ **Process selection rules**

/* Processes are selected by matching the required operations with available capabilities. */

- Rim requires turning for OD (outer diameter), turning for ID (inner diameter), face-milling operations (both roughing and finishing operations) and grinding.
- Web requires turning the ID, milling web contour, then grinding.
- Hub needs turning of ID, face milling, then grinding.
- Airholes can be milled or drilled.

☐ **Operation sequencing rules**

/* These decision rules specify an operation sequence or process plan. */

- First machine the hub, to use as a supporting surface for other operations.
- Next, start from outermost feature (that is, rim), and move towards the hub.
- Machine both sides of the disk simultaneously where possible.
- Finish all operations which can be executed on one machine before proceeding to another.
- Follow a roughing cut by a finishing pass.
- Grind a finished part to final dimensions.

■ Appendix B: sample interaction

A session using the turbine disk advisor is illustrated by the following dialogue. User input is in **bold**, while meta-level comments are denoted by the symbols /* and */.

|?– turbine_disk_advisor.
What is the operating temperature (in C) of the disk?
|: **it's 2500 C.**
What is the operating speed (in rpm) of the disk?
|: **about 2000.**
What is the weight (in lb) of an individual blade to be arranged to the disk?
|: **about 6 lb.**
What is the desired rms surface finish (in micrometres) of the slots?
|: **90.**
What is the tolerance (±1000ths of an inch) of the slots?
|: **say 10.**
What is the desired rms surface finish (in micrometres) of the web?
|: **100.**
What is the tolerance (±1000ths of an inch) of the web?
|: **10.**
What is the outer diameter (in 26ths of an inch) of the hub?
|: **64.**

> ●
> ●
> ●

/*   The dimensions, tolerances and surface finishes of each of the other features are then requested. These features relate to the web, rim and airholes. */

/*   For this simple interaction, the system starts off with a design space containing three designs. The system presents all tractable solutions for final selection by the user. */

Solution 1 is as follows:

outer diameter:	80 (16ths of an inch)
inner diameter:	64 (16ths of an inch)
width:	60 (16ths of an inch)
surface finish:	100 (micrometres)
tolerance:	10 (1000ths of an inch)

/*   The specifications of the remaining features of design #1 follow. */

> ●
> ●
> ●

/*   An informal sequence of operations along with the machines is generated for design 1. This sequence includes operations such as turning, facing, drilling, broaching and grinding. */

Sequence of operations:

operation:	Turning_inner_diameter_of_hub.
Machine name:	Lathe 1
operation:	Turning_outer_diameter_of_hub.
Machine name:	Lathe 1
operation:	Turning_inner_diameter_of_rim.
Machine name:	Lathe 1

•
•
•

/*   The output for designs 2 and 3 follow in similar fashion. */

4.3

Production-oriented design: a knowledge-based approach

K. G. Swift,
M. E. Uddin,
M. G. Limage and
M. S. Bielby

Department of Engineering Design and Manufacture, University of Hull, UK

■ Introduction

The success of a product in the market place is dictated by its costs, performance and reliability, which are predetermined to a very large extent by the work of the designer. In fact the design of a product largely defines how it is to be made and what it will cost to produce. The designer's problem is, therefore, to create products which not only satisfy the functional requirements, but are economic to manufacture. Thus he or she needs to have expertise in a wide range of fields, including specialized topics such as materials, manufacturing and assembly technologies. It may take many years to accumulate all the knowledge required.

The high cost associated with changes in design, once production has begun, means it is vitally important to make the correct design decisions at the drafting stage. The requirement is for the timely provision of technological and economic data enabling the designer to make detailed assessments of alternative product configurations, manufacturing and assembly costs throughout the design process.

This paper is reprinted by permission of the authors and Butterworth & Co. (Publishers) Ltd from *Advanced Manufacturing Engineering*, Vol. 1, January 1989

This paper provides a summary of research work being done at the University of Hull, which arose from the idea that knowledge-based systems or expert systems, used in conjunction with a CAD workstation, might offer a solution to such design problems. A knowledge-based system offers a convenient method of presenting the information required allowing the designer to take better decisions with greater confidence. The paper concentrates on the problems of providing the designer with a means of quantitatively comparing alternative design solutions for assembly and component manufacturing costs and discusses the application of producibility knowledge in the automated drafting process.

■ A technique for design evaluation

Analysis of the design of products for manufacture and assembly processes is basically done by using the procedure shown in Figure 1. The analysis is an iterative process involving straightforward procedures including functional analysis, manufacturing analysis, and evaluation of feeding, gripping and fitting processes.

Before product design begins it is important to decide whether each product is unique, or whether there are similarities, and therefore if there is an

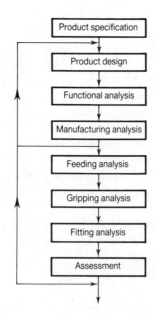

Figure 1 Design evaluation procedure.

Why assemble? The points below illustrate the need:

1 **Degrees of freedom (movement)**: Various elements must enjoy a degree of mobility in order to achieve the function

2 **Material differentiation**: The function's realization depends on particular material characteristics

3 **Production considerations**: Some parts will be easier to produce by division into subparts

4 **Establishment condiderations (replaceability)**: The product may be used in a fixed installation and have to be assembled in a separate process

5 **Differentiation of functions**: A function can be carried out by a single agent or a combination of such in a form of more elements

6 **Particular functional conditions**: In the sense of increased requirements of accessibility, demounting, cleansing, inspection, etc; these can necessitate a division into elements

7 **Design considerations**: Aesthetic requirements can cause a division of the form which will consequently require assembly

Figure 2 Reasons for assembly.

opportunity for standardization of components and assembly procedures, and for establishing product family themes. This is a strategic issue with many implications including the viability of automation equipment. In the functional and manufacturing analysis stages we are basically trying to assist the process of design rationalization.

Assembly costs may be reduced by the elimination of components or their combination with others. Criteria have been developed for assisting this process broadly based on the factors shown in Figure 2, taken from Andreasen *et al.* (1988). The functional analysis stage aids the designer in the identification of any components that are candidates for elimination or combination with other parts and enables the calculation of a design efficiency (percentage) for the product or assembly based on the minimum possible number of essential components to the total number of components in the assembly. The analysis encourages the creation of rationalized assembly design solutions (Figure 3).

A consequence of this functional analysis is that it is possible to deflect the problems of assembly by designing a few complex components instead of many simple ones. Clearly, the cost for manufacture of complex components could possibly erode any advantage gained in reduction of assembly costs. To this end the manufacturing analysis can be used to determine a comparative cost for component manufacture.

A goal of the manufacturing analysis is to enable the designer to anticipate the consequences on manufacturing costs of assembly design rationalization and design changes for ease of handling and assembly. We

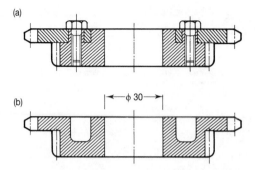

Figure 3   Two different types of combined sprocket and gear wheel: (a) includes assembly: this is made of steel and produced by machining. The individual teeth are cut. It is necessary to divide it into two elements: in a nutshell, assembly due to production technique reasons; (b) does not include assembly: this is produced from sinter metal, the teeth being sintered in accordance with the required tolerance and surface quality; the advantages being no waste material, short processing time and no assembly.

have found that the absence of such a facility is one of the major reasons for the limited adoption of design for assembly, DFA, techniques.

In order to calculate the cost of manufacture of one design of component against another, manufacturability properties or parameters are quantified for each component and used as cost coefficients in equations for cost enumeration. The coefficients modify the elemental cost including overheads, etc., of producing the ideal component design (material, features, etc.) by a certain production process at a particular quantity level. In this way the cost of manufacture can be determined for any component, whether produced by one or more processes, which can be represented by a series of cost coefficients. The basic equation used to provide an indication of component costs is:

$$\text{Manufacturing cost } (M) \propto V C_{mt} + R_{cc} P_c$$

where V is the volume of material needed to produce the finished component, C_{mt} is the volumetric cost of material in required form, P_c is the basic cost of producing the ideal design of component by a specific process and R_{cc} is the relative cost coefficient assigned to a component design, taking account of shape complexity, minimum section, suitability of material for process, tolerances and surface finish.

The technological and economic knowledge comes from experts in companies specializing in particular production processes. Currently the manufacturing analysis considers processes including: plastic moulding, die casting, machining (capstan and auto), powder metallurgy, impact extrusion, forging, press working and sand casting techniques. The analysis is restricted to small components only. Initial trials have shown that the technique may be capable of predicting component costs in the early stage of design to an accuracy of around 10%.

The feeding analysis is concerned with the problems of handling components and subassemblies from the point of manufacture to the point of presentation within the assembly system. The feeding analysis is made on each component in three stages:

- Stage 1 examines the possibility of transporting the component from the point of manufacture to the assembly system in an orientation-maintained manner.
- Stage 2 assesses the general physical properties for feeding of those components that cannot be transported with orientation maintained.
- Stage 3 examines the suitability and practicability of component designs for feeding.

The object of these stages is to ensure that the design of individual components is compatible with the likely method of feeding. Those components which are either well suited for retained orientation or straightforward feeding methods score lowest cost indices.

The gripping analysis examines the ease with which each component can be held for transportation from the point of presentation within the assembly system to the stage where insertion begins. Each component is examined for its suitability for gripping and is assigned the appropriate gripping index.

The fitting process analysis requires the designer to generate an assembly sequence flowchart and to assign a cost index to the individual processes. The analysis cannot be made unless a proposed sequence of assembling the product is declared. The analysis is made on each individual insertion operation involved in production or assembly construction. The objective of the fitting process analysis is to identify fitting processes that are expensive. High individual relative cost values and/or a high total of individuals indicate costly assembly processes and, therefore, suggest redesign considerations with a view to minimize the fitting (placing, workholding and fastening) costs. The way in which the evaluation procedure influences the costs of assembly can be explained by considering the simple relationship:

$$\text{Assembly cost} \propto \sum^{N_p} (C_h + C_g + C_f)$$

where N_p is the number of parts in assembly, C_h is the handling cost of an individual part, C_g is the gripping cost of an individual part and C_f is the fitting cost of an individual part. The valuation procedure produces a manufacturability status report, MSR, for the product or assembly. The MSRs can then be used as an aid to generate improved proposals which are likewise assessed. The procedure has been developed in close collaboration with Lucas Industries plc who evaluated it internally over a number of years before adopting it as their standard for assessing new designs and existing products for manufacturability (Miles and Swift, 1988). Also, Lucas offer a consultancy service, based on this technique, to other organizations. The application of this evaluation is illustrated in the next section of the paper.

■ An illustrative example of the evaluation procedure

The application of the design evaluation procedure to two differently designed staplers, is described here (Figures 4 and 5). These figures illustrate the two designs: both employ the same working principle, deliver the same class of staple (No. 10) and originate in Japan. Their MSRs, presented in part in Figures 6 and 7, show significant differences between the two designs in

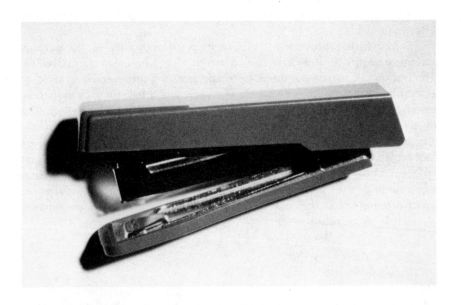

Figure 4 Stapler design A.

Figure 5   Stapler design B.

terms of assembly structure and in all the analysis stages. (The results of the evaluation for Figures 4 and 5 are shown in Figures 6 and 7, respectively.) The main differences are summarized in Table 1. The results presented in Table 1 are now briefly discussed.

☐   Design efficiency

These results show that even with a well-known working principle considerable variation in design can exist. The reasons behind the choice of configuration are complex, but the analysis exposes opportunities for rationalization. Design efficiency values of less than 50% are generally considered to indicate potential for rationalization.

☐   Manufacturing cost

The total manufacturing values represent the estimated cost of producing all the components in the design. While the results shown are estimated cost (pence), more confidence exists in their relativity knowing the problems of

Component number	1	2	3	4	5	6	7	8	9	10	11
Functional analysis	B	B	A	A	B	B	B	B	A	A	B
Manufacturing analysis	1.52	1.09	1.28	1.83	1.07	1.24	1.73	0.63	1.50	1.07	0.75
Feeding analysis	1.1	1.3	1.1	4	1.3	2.5	1.3	1.2	12	1.3	1

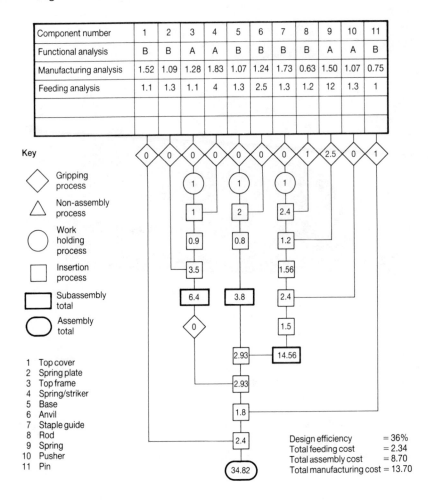

Key

◇ Gripping process

△ Non-assembly process

○ Work holding process

□ Insertion process

▭ Subassembly total

⬭ Assembly total

1   Top cover
2   Spring plate
3   Top frame
4   Spring/striker
5   Base
6   Anvil
7   Staple guide
8   Rod
9   Spring
10 Pusher
11 Pin

Design efficiency   = 36%
Total feeding cost   = 2.34
Total assembly cost   = 8.70
Total manufacturing cost = 13.70

Figure 6   Part of MSR for stapler A.

industrial rate-fixing practices. The results indicate that the components in design A are 44% more costly than in design B.

☐   **Feeding cost**

The total feeding values suggest that once again there are cost benefits in design B. (The total feeding cost, in pence, is derived by assuming the blend of feeding technology identified as most appropriate during the analysis process.) The costs of handling the components in design A are 15% greater than that of design B.

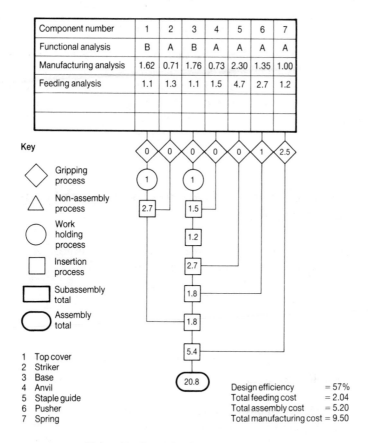

Component number	1	2	3	4	5	6	7
Functional analysis	B	A	B	A	A	A	A
Manufacturing analysis	1.62	0.71	1.76	0.73	2.30	1.35	1.00
Feeding analysis	1.1	1.3	1.1	1.5	4.7	2.7	1.2

Key

◇ Gripping process

△ Non-assembly process

○ Work holding process

▭ Insertion process

▢ Subassembly total

⬭ Assembly total

1 Top cover
2 Striker
3 Base
4 Anvil
5 Staple guide
6 Pusher
7 Spring

Design efficiency = 57%
Total feeding cost = 2.04
Total assembly cost = 5.20
Total manufacturing cost = 9.50

Figure 7   Part of MSR for stapler B.

Table 1   Summary of analysis results.

Parameters	Stapler A	Stapler B
Total number of parts	11	7
Minimum possible number of parts	4	4
Component design efficiency (%)	36	57
Total feeding cost (pence)	2.34	2.04
Total fitting cost (pence)	8.70	5.20
Total manufacturing cost (pence)	13.70	9.50
Overall production cost (pence)	24.74	16.74
Annual production	10★	10★

★From Spivey (1984).

☐ **Fitting cost**

The total fitting values indicate that there are large differences in the costs of product construction between the two designs. Design A is 67% more costly to assemble than design B. The fitting cost estimate also relies upon assumptions regarding the assembly technology employed. These, and the assumptions regarding feeding, tend to be the areas of main concern in these cost predictions and are topics of current investigation.

■ Computer-based design for assembly and manufacture

The computer may be used to assist the analyst in a number of ways. The benefits compared with the manual application include:

- paperless operation;
- less tedious and time saving;
- consistency of results and their presentation;
- more knowledge may be used without overpowering the user;
- possibility for automatic evaluation and redesign in CAD.

Computer-aided design for assembly can be pursued by the development of conventional programs (Boothroyd and Dewhurst, 1987); however, since the solution of design for assembly and manufacture problems basically requires a large amount of detailed knowledge, experience and judgement, a knowledge-based approach (Winston, 1984; Michie, 1982) to the field was adopted. A rule-based expert system has been developed having facilities including:

- advice on the consequences of design decisions on assembly costs and suggestions for redesign;
- explanation of questions and deductions, and assistance with terminology;
- some treatment of uncertainty in user responses and rule validity;
- satisfaction of rule conditions and redesign propositions of the knowledge base through its integration in CAD.

The main elements of the system are illustrated in Figure 8. The rule interpreter is coded in PROLOG (Spivey, 1984; Bonsai, 1987) and designed to investigate a collection of PROLOG clauses (rules) in the satisfaction of system conclusions by using mostly backward chaining. (The system

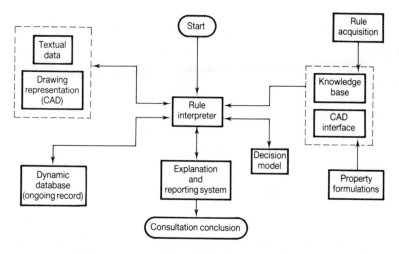

Figure 8 Overview of the consultation system.

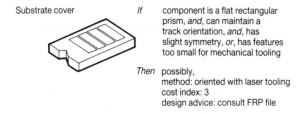

Figure 9 A sample rule.

currently has around 200 system goals or conclusions and these are investigated using approximately 600 inference rules.) An example rule taken from the knowledge base is shown in Figure 9. The system knowledge is transparent and readily displayed, accessed and modified. More detail on the representation of knowledge used and its interpretation can be found in Swift (1987).

The expert system has been implemented on a PC where the designer is supported through a conversational dialogue in the making of the evaluation procedure overviewed earlier. The PC version has been developed over a number of years and is available commercially (Lucas).

■ Application in CAD

At present most of our research effort in the area of computer-based design analysis is focused on the application of the evaluation procedure in CAD. At the lowest level, application in CAD could be achieved merely by using the expert system, in a wholly conversational mode, as a process in the CAD workstation. However, since the majority of rule conditions in the knowledge base may be satisfied by geometric data, it is worth exploring the opportunities for automatic analysis of drawings as they are produced. In addition we have developed software that can be used to provide assistance in the preparation of the MRS. The software enables the designer to rapidly generate the assembly sequence flowchart and automatically places the evaluation indices in their appropriate locations following their determination by the knowledge-based system.

☐ Geometric reasoning

Work under this heading has been concerned with two main areas: design of components to suit automatic handling processes and design for ease of assembly. In both of these, consideration has been given to automatic determination of properties that influence either handling or fitting directly from design drawings and also the communication of design revisions by the redrawing of inappropriate designs.

☐ Determination of properties from drawings

A vital issue in this connection is the provision of a representation of drawings that is adequate for use by an expert system in the automatic determination of salient properties. It should be noted that the intention here is not to propose an alternative representation for drawings in CAD systems, rather the approach is one of using as straightforward a representation as possible that can be easily derived from whatever representation is already available.

Consider components that are rotationally symmetric. For such components handling and assembly properties can be determined from a rotational section defined by lists of (X, Y) nodes and internodal surface attributes. The attributes completely describe the geometry between the nodes. By using such a representation we know that properties relevant to component handling can be automatically extracted from drawings generated in CAD (Swift, 1987). This approach is also being used to investigate the assembly of components (Figure 10). Component assembly also requires the investigation of the union of the co-ordinate and attribute lists in order to define the relevant insertion properties.

When considering component placing and fastening a number of design-related properties are important:

- direction of component access (single/multiple axis);
- ease of alignment and positioning;

- resistance to insertion;
- stability of components once inserted/assembled;
- fastening technology.

To get a feel for what can be done let us touch the area of ease of alignment and positioning.

In general components are easy to align and position during assembly if the co–ordinate and attribute lists, defining the union of the components, contain geometric features (apt, chamfers, leads, etc.) at the relevant locations.

Such definitions can be readily formulated as PROLOG clauses in terms of our structured sets of lists. A sample PROLOG formulation involved in defining ease of alignment and positioning may be found in Figure 11.

☐ Communication of revisions by redrawing

A distinctive requirement of our expert system is the ability to identify variations in the ground data (the component design) that would lead to a major reduction in handling and assembly difficulty. This may be contrasted with other expert system applications such as the interpretation of medical systems or geological phenomena, where the data cannot be varied.

Design advice is issued where handling and assembly processes are not viable. The advice can be presented simply as a descriptive text retrieved from a rule of the knowledge base. A more interesting approach might be to communicate suggested design revisions by displaying possible modifications to the drawings produced. This has the benefit of letting the designer

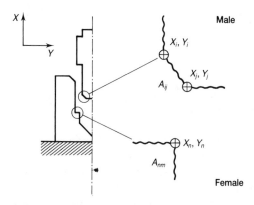

Figure 10 Representation of component fitting process: X, co-ordinates in the x-direction; Y, co-ordinates in the y-direction; A, internode surface attributes.

```
alignment and positioning (Index):-
      has male alignment feature (Xm, Ym, Am, Featurem),
      has female alignment feature (Xf, Yf, Af, Featuref),
      alignment index (Featurem, Featuref, Index).

has male alignment feature (Xm, Ym, Am, chamfer):-
      has chamfered surface (Xm, Ym, Am),
      angle of chamfer (Xm, Ym), size of chamfer (Xm, Ym),
      position of feature (Xm, Ym), correct orientation (Xm, Ym).

has male alignment feature (Xm, Ym, Am, roundend) . . .

correct orientation (Xm, Ym):-
      guides part into position (Xm, Ym).

size of chamfer (Xm, Ym):-
      chamfer large enough to influence assembly (Xm, Ym).

angle of chamfer (Xm, Ym) . . .

alignment index (chamfer, nofeature, 1.2).

alignment index (chamfer, chamfer, 1.0).
```

Figure 11   Defining alignment and position.

know what a solution to the problem could look like. It should be noted that we are in a situation where it is unlikely that any proposed design solution will be unique, and the best choice may not only depend upon the particular features of the component, but also on manufacturing aspects and functional constraints not fully represented in the system.

Some moves have been made in this direction to the extent that non-functional geometric features can be added to component drawings generated in CAD for the purposes of eliminating automatic orientation problems (Swift, 1987). Redrawing has been achieved by modifying the representation used to describe the properties of the current design, to include the orientation features required at the appropriate locations in the co-ordinate cycle. The techniques developed for redrawing are based upon the powerful list-processing possibilities in PROLOG. Work is currently focused on the idea of redrafting of assembly designs to facilitate insertion processes.

■ Industrial application

The PC version of the expert system has been used in Lucas Industries plc on DFA applications. During these studies it has been possible to take a few measurements with regard to the performance of the expert system.

The performance measurements have shown that the software can promote the generation of consistent results and can offer a time saving of up to 30% compared with purely manual evaluation procedures. The time taken to make an analysis clearly varies from product to product. In one series of trials five different single users were able to complete an analysis on a fairly complex 20 part-roller ball assembly in times varying between two and three hours by using the PC software.

■ Concluding remarks

Application of the manufacturability evaluation methodology outlined has shown that it can be used to identify technological and economic weaknesses in product assembly and manufacture. Such evaluations should be included as a standard part of drawing office procedures.

A knowledge-based expert system can be used to enhance the application of the evaluation procedure and its integration in CAD should further aid the practice of these techniques in a manufacturing organization.

References

Andreasen M. M. *et al.* (1988). *Design for Assembly* 2nd edn, Chapter 2. UK: IFS Publications; Berlin: Springer-Verlag

Bonsai (1987). *Arity PROLOG*. Bonsai Limited, 112–116 New Oxford Street, London

Boothroyd G. and Dewhurst P. (1987). *Product Design for Assembly*. Boothroyd Dewhurst Inc., Wakefield, RI, USA

Lucas Engineering and Systems Ltd, Lucas Factory Systems, Solihull, West Midlands, UK

Michie D. (ed.) (1982). *Introductory Reading in Expert Systems* Vol. 1, Part 1. London: Gordon and Breach Science Publishers

Miles B. L. and Swift K. G. (1988). Design for assembly. *Automotive Engineer* (June/July)

Spivey M. (1984). *University of York Portable PROLOG System* Release 2. Department of Computer Science, York

Swift K. G. (1987). *Knowledge-Based Design for Manufacture* Chapter 6. London: Kogan Page

Winston P. H. (1984). *Artificial Intelligence* 2nd edn, Chapter 6. Reading, MA: Addison-Wesley

Acknowledgements

The work described in this paper is supported by the ACME Directorate of the SERC, Grant Number: GR/E 02949.

The authors thank Mr B. L. Miles, Mr G. Hird of Lucas Engineering and Systems, Mr D. Nunn of D.N. Automation and Mr B. Barmby of Armstrong Fastening Systems for their invaluable assistance and encouragement.

4.4
ASSYST: a consultation system for the integration of product and assembly system design

**F. Arpino and
R. Groppetti**

Politecnico di Milano, Italy

■ Introduction

Product design and redesign for easy assemblability is a very critical task, particularly when automatic flexible assembly has to be introduced into the factory. As known the competitiveness of the product is mainly based on product specification and differentiation and on production costs reduction (Takahashi and Senba, 1986). Therefore it is required, starting from the early product design stage, to consider product in relation with assembly process and assembly system, particularly with the assembly technology and level of automation, the structure or configuration of the assembly system, the selection of the assembly equipment (Andreasen and Ahm, 1986). This problem has been solved generally by means of the so-called design for assembly methodologies and tools, applied at the early design stage. Several methods and tools for design for assembly have been presented in the literature (Boothroyd et al., 1978 and 1982; Boothroyd and Dewhurst, 1983; Dwivedi and Klein, 1986; Eversheim et al., 1986; Graves and Poli, 1984; Jakiela et al., 1985; Poli and Graves, 1985; Poli et al., 1986; Poli and Fenoglio,

This paper is reprinted by permission of the authors and IFS Ltd from *Developments in Assembly Automation*, March 1988, pp. 167–80

1986; Ramsli and Neerland, 1986; Swift and Finth, 1984; Takahashi and Senba, 1986). Some of them are collections of advice or handbooks and others are software packages. They are based on the evaluation of a set of product attributes and identified cost drivers, in order to facilitate assembly operations, these being manual or automatic, considering a predefined technology, specific devices and generally a predefined one-kind alternative assembly system (manual or automatic dedicated or automatic programmable flexible). This approach limits the possibilities to evaluate all the assembly system alternatives in a systematic and easy way, and to evaluate also all the implications and consequences of each product design choice on assembly system design, at each step of the early product design/redesign iterative decision-making process. This approach generally allows assembly cost and time estimation, with the beforesaid limitations, but it could lead to suboptimized solutions for product, during early design or redesign, and later for the assembly system, during preliminary assembly system design, and, if an optimized global solution is searched for, it could not satisfy time to market requirement. Because product and assembly system have to match in order to obtain production costs reduction, there is the need of an early interrelated design that leads to a global solution product/assembly system (Andreasen and Ahm, 1986; Withney et al., 1984).

Globally optimized solutions for the whole design process could be obtained and anticipated at the product design stage only by means of an integrated and systematic approach: assembly systems should be preliminarily defined and evaluated during product analysis for ease of assembly, in order to design a product just right for the best selected assembly system.

The need of an integrated solution is particularly strong in a FAS/CIM, flexible assembly system/computer-integrated manufacturing, environment, where product definition data, feasible assembly operations data, alternative assembly systems data and available assembly resources data have to be communicated or shared among factory functions. This integrated solution has to be searched for the following typical situations:

- new product to be adapted to an existing assembly system or equipment;
- new product and related new assembly system to be designed;
- existing product to be redesigned for a new assembly system.

A true integrated approach, in principle, should integrate assembly process into the whole manufacturing process. This means that assembly product design requires the analysis of component design, not only regarding assembly operations but also all the other technological processes (machining, injection moulding, . . .). This analysis is based on the conciliation of product assembly requirement with component manufacturing requirements. But, even if some 'design for producibility' methodologies are in development, this multi-process global approach still presents some difficulties.

Among the different proposed design methodologies and tools, we have followed an integrated approach implementing a prototype consultation system, named ASSYST (ASsembly SYSTem), that offers integrated tools for product assemblability evaluation, with respect to alternative assembly system configurations, and for preliminary assembly system design. For each initial product definition and production requirement, ASSYST provides automatically the evaluation of the product design for ease of assembly and, if required, some advice for its redesign. At the same time and for each product configuration, ASSYST provides automatically a set of basic feasible alternative assembly systems (automatic programmable flexible single/multi-station, automatic dedicated synchronous/asynchronous, manual single/multi-station), that are specified in terms of structure, devices for transport, feeding/orienting, insertion, etc. If required, designer can modify interactively these basic systems assigning a specific device type for each operation. Then ASSYST evaluates automatically the capabilities of the systems and provides a complete cost and investment analysis, that allows the rational and systematic comparison among alternatives and the selection of a globally optimized and integrated product/assembly system solution. In order to exemplify the methodology and the operation of ASSYST and the benefits from its application to the redesign of industrial products and to the preliminary design of the relevant assembly systems it can be advisable to present and discuss also a simple case study.

■ Methodology and architecture

ASSYST system is a prototype of an integrated tool, at product/assembly system early design stage, for design for assembly and assembly process analysis. The methodology followed for product analysis can be summarized as follows (Figure 1). The first step requires the definition of the product as a structure of subassemblies, components, listed as in a bill of material, and alternatives among them, corresponding to different product variants. Subsequently each component will be described in terms of assembly form features and attributes, particularly for feeding and orienting operations. The connection logic and the assembly sequence are described as a set of inserting and securing operations applied to each component in order to make it a product part.

After product definition phase, all the assembly attributes are evaluated with respect to different assembly technologies that can be applied. For a simplification hypothesis, these are grouped into: manual assembly, dedicated automatic assembly, programmable flexible automatic assembly. By the evaluation of the influence of each attribute on the assembly process the

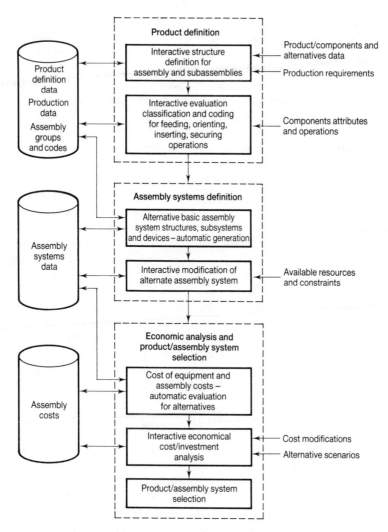

Figure 1   ASSYST methodology and architecture.

assembly operation time, the device necessary, its capabilities and cost in the specific case are automatically determined. From this analysis for each component the assembly times for manual assembly operations (grasping, orienting, inserting, securing), the evaluation of type, capacity, cost of devices needed for feeding and orienting, inserting and securing/fastening for dedicated automatic assembly, the complexity and costs of needed devices and operation time for programmable flexible automatic assembly can be obtained. As a result it is available as a set of basic elements or assembly devices, for each operation related to each component and for each alternative assembly technology. These basic elements can be combined to define a set of alternative assembly system structures. System definition is essentially based on the selection of the devices required for each assembly operation within available alternative assembly technologies; it is also required for this definition to decide annual production volume, cycle time and the consequent assembly tasks distribution among all the different workstations. Moreover, when more operations can be executed by only one device, as in manual and programmable assembly, also single-station systems can be defined. If a material handling/transport subsystem and workcarriers are required the related cost can be evaluated (Funk, 1986; Poli and Groppetti, 1986; Poli et al., 1986; Poli and Fenoglio, 1986; Ramsli and Neerland, 1986). Therefore assembly costs and investment costs of assembly equipment can be easily evaluated (Poli and Groppetti, 1986) for each modelled system. By means of suitable investment analysis methodologies (Crumpton, 1980; Czajkiewicz, 1986) it is easy to evaluate the economic advantage of a certain product/assembly system solution. Hence the developed tools are suitable to evaluate design modification, or redesign actions, suggested by design for assembly methodologies and by design for assembly knowledge and practice, both to product structure and components attributes, and to assembly system structure and to the combination of assembly devices. The iteration of the analysis and decision-making process make easy the right quantitative evaluation, with a good degree of accuracy, of all the effects of these modifications on product and assembly process. This iteration makes easy to get results good enough in a reasonably short time, giving an integrated and holistic view of product and assembly system.

Because the major point of processing operations, needed for the realization of the analysis tool, is made by processing of a conspicuous amount of data from different sources, for which is required an organic and structured representation, the system has been based on DATA-TRIEVE, a database management system by DEC, implemented on a VAX 11/750 under VMS v4.5 operating system. In order to improve the user interface FMS, form management system, also by DEC, was applied. As shown in Figure 1, the architecture of the tool can be subdivided into two fundamental subsystems: the database and the application modules. In the database the fundamental elements are the partitions of product and systems data; moreover other areas are related to classification and coding and to other assembly costs, as operators' costs, annual working hours, etc. The application modules are the elements executing the required transformations and data processing, and that makes possible the interaction with the designer. It is possible to identify three basic application modules:

- product definition module, that contains a submodule for interactive product structure definition for assembly and subassemblies, a submodule for classification coding and attributes evaluation for feeding, orienting, inserting, securing/fastening operations;
- assembly system module, made by two submodules for automatic

generation of alternative basic assembly system structures and configuration, with the definition of subsystems and devices and for the
interactive modification of these alternative assembly systems;

- economical analysis and product/assembly system selection module,
  made by two modules for investment cost of assembly equipment and
  assembly costs automatic evaluation for each alternative, interactive
  economical cost/investment analysis for alternative scenarios and
  product/assembly system selection.

■ Data representation structure

The data structure applied for both product and assembly systems models
representation can be illustrated as follows.

□ Product data

The most part of information regarding real product is stored, because
some product features and specifications, for example, the existence of
more than one product variant, have a great influence on product/assembly
system study. Moreover also product subassemblies were considered in
product representation.

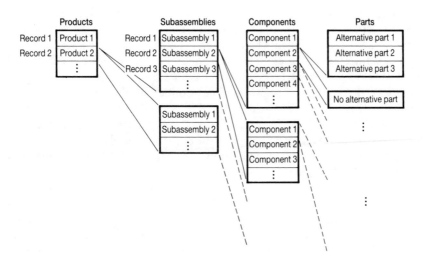

Figure 2   Product database structure.

Adopted data representation is based on a hierarchical record structure (Figure 2). The first level is made by product database, where product identification code, product name, requested production volume, number of variants, etc. are stored. The second level stores information regarding identified subassemblies. The whole product, by itself, is considered as a subassembly; consequently product information is stored at this level. Data considered here are: product id code, to identify product to which are related, subassembly id code, subassembly level (0 for assembly, 1 for the first level subassembly, and so on), subassembly name, quantity, number of variants. The third level is represented by the list of components for each subassembly; therefore the main pieces of information are: product, subassembly and component id codes, component name, quantity, order number in assembly sequence, alternative number, that is, the number of alternative parts. These parts represent alternatives of components for variants of a certain product; therefore components unify variants at a conceptual level. The fourth and last level is assigned to store attributes of parts features. At this level only information related to assembly attributes is stored, because the previous levels are essentially assigned to represent assembly product structure.

Data stored at the fourth level are relevant to both feeding/orienting attributes, subdivided between rotational and non-rotational or prismatic, and inserting/securing operations. By means of this scheme it is possible to represent many products, each one with its own complete subassembly structure, taking in consideration at the same time more than one product variants presence.

☐ **Assembly systems data**

Assembly systems data also are structured in a hierarchical system (Figure 3). The lowest level is component assembly. This is related, record by record biunivocally, to components level inside product database. Basically device typology for feeding, orienting, inserting and securing and the number of these devices, requested to satisfy variants requirements, are stored; moreover there are data regarding the number of alternative stations, if required, to assemble alternative parts. These last data are stored also at the upper level, represented by stations. At this level the devices inside assembly stations will be evaluated, either dedicated to a single component (for example, feeders), or to all the operations inside the station (for example, data are stored a datum regarding the utilization of only one robot inside the station, although more than one component is assembled). Moreover the number of effective stations and total operation time are stored. The upper level is represented by the complete assembly system; at this level, total number of stations in the system, total equipment cost, direct operators number, indirect operators and required supervisors number, single-system volume capacity, number of systems needed in parallel, etc. are stored. Each

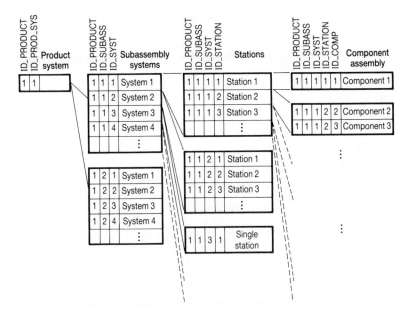

Figure 3   Assembly systems database structure.

record at the upper level can contain a list, of variable length according to the number of defined subassembly; in this list is coded the selected assembly system for each subassembly. If more records are built, it is possible to analyse different assembly system solutions, for the same subassembly defined in the product. The use of this further level makes possible product evaluation without the limit of a subassembly only at each time and to analyse easily different system combination alternatives.

■ Fundamental modules and operation

The procedures allowing interactive data modification in the database and automatic execution of all the tasks for product/assembly systems representation and analysis have been conceptually grouped into three separate modules. The first module to be used allows assembly product definition, from its structure with subassemblies, components and variants, down to assembly attributes and inserting operations. This last information, strictly related to components, is analysed automatically by the classification and coding submodule, that allows, as regards feeding and orienting operations, to obtain in output a code, as used in Boothroyd *et al.* (1978). That allows the search for and the selection of suitable orienting devices.

The configuration and modification module of assembly system models allows the automatic definition of a set of fundamental system architectures, belonging to the fundamental assembly technologies (manual, automatic dedicated, automatic programmable flexible assembly):

- manual single station;
- manual multi-station;
- manual multi-station with automatic feeders;
- automatic dedicated with stacks;
- automatic dedicated with feeders;
- automatic programmable flexible single station with stacks;
- automatic programmable flexible single station with feeders;
- automatic programmable flexible multi-station with stacks;
- automatic programmable flexible multi-station with feeders.

For each assembly system configuration the equipment cost of investment is evaluated for the whole system, for each station, down to each feeding/orienting device. Then line balancing, if necessary, and production volume capacity for each assembly system, and the number of systems required in parallel, if that is the case, to satisfy total requested volume, are estimated.

Assembly systems can be evaluated by the economical module in the obtained condition or they can be modified interactively in order to satisfy in a better way requirements related both to product design specifications and generally to production. In fact it is possible to modify the selected feeding/orienting or inserting/securing device for each component, inside each or inside all the systems, to modify also cycle time or the number of shifts, etc.: then ASSYST will evaluate again automatically the new modified solution.

The module for economical evaluation is a very suitable and versatile tool for investment analysis. It is based on a software spreadsheet programmed for DCF, discounted cash flow, analysis. ASSYST makes easy and fast the use of the spreadsheet because it provides automatically the main data required for the analysis, as total investment cost, number of operators, number of shifts, production volumes, energy costs, interest rate, wages, etc.

One spreadsheet can be used for each system at subassembly level, or it is possible to use it for the evaluation at assembly product level. The spreadsheet allows also to evaluate NPV, net present value, and, defining other production costs and sales, it allows a complete analysis and to evaluate NPV, IRR, internal rate of return, payback period, etc. Each value, entered by the designer or automatically by the system, can be modified and, if it is not exclusively pertaining to a certain assembly system, affects automatically the evaluation of all the systems. Therefore it is easy to iterate the analysis several times, as required, in order to evaluate the influence of the considered parameters with different scenarios and assembly system alternatives, and to introduce modifications at the product design stage in order to select not only a product designed for ease of assemblability but an optimal integrated product/assembly system solution.

■ Conclusion

An integrated approach to product/assembly system design to reach a globally optimized product/assembly system solution, requires, as shown, from the early design stage, to analyse product with respect not only to assembly operations but also to assembly system alternatives, and possibly considering also all the other technological processes relevant to product components. The prototype consultation system presented and the case study discussed have shown the feasibility of such an approach, to follow an integrated product/system design for assembly and redesign methodology, even if a system based on a fully integrated multi-process global approach is still not available and at present the system suggests only implicitly advice for product redesign for ease of assembly, at the same time, with an estimate of the saving resulting from the adoption of this redesign and modification advice.

Further research is in progress in the area of product representation and definition by means of a CAD 2D and 3D modeller with an assembly attribute and form feature recognizer, in the area of the integration of these modules with the consultation module for product/assembly system design, in the area of the relational approach for product and production system data representation and finally in the area of artificial intelligence techniques for the representation and activation of design for ease of assembly rules.

References

Andreasen M. M. and Ahm T. (1986). The relation between product design, production, layout and flexibility. In *Proceedings of the 7th ICAA*, Zurich, IFS (Publications) Ltd

Behuniac J. A. (1984) Design for assembly automation: the competitive edge. In *Proceedings of the 5th ICAA*, Paris, IFS (Publications) Ltd

Boothroyd G. and Dewhurst P. (1983). *Design For Assembly Handbook*. Amherst: University of Massachusetts

Boothroyd G., Poli C. and Murch L. E. (1982). *Automatic Assembly*. New York: Marcel Dekker Inc

Boothroyd G., Poli C. and Murch L. E. (1978). *Handbook of Feeding and Orienting Techniques for Small Parts*. Amherst: University of Massachusetts

van Brussel H. and De Winter D. (1984). Computer aided design of flexible assembly systems. *CIRP Annals*, **34**(1)

Crumpton P. A. (1980). The economic justification of automatic assembly. In *Proc. of the 1st ICAA*, Brighton, IFS (Publications) Ltd

Csakvary T. (1981). Product selection procedure for programmable automatic assembly techniques. In *Proceedings of the 2nd ICAA*, Brighton, IFS (Publications) Ltd

Czajkiewicz Z. J. (1986). Justification of the robots applications. In *Flexible Manufacturing Systems: Methods and Studies*. North-Holland: Elsevier Science Publishers

Dewhurst P. and Boothroyd G. (1986). Computer analysis of product design for robot assembly. In *Acts of International Conference on Computer-Aided Production Engineering*, Edinburgh

Dwivedi S. M. and Klein B. R. (1986). Design for manufacturability makes dollars and sense. *CIM Review*, Spring 1986

Eversheim W. *et al.* (1986). Montagegerechtes Konstruiren mit CAD. *Industrie-Anzeiger*, **108**(20)

Funk J. L. (1986). The potential market for robotic assembly. *International Journal of Production Research*, **24**(3)

Graves R. J. and Poli C. (1984). Integrated product design and assembly process design. In *Industrial Engineering Conference Proceedings*, Institute of Industrial Engineers

Jakiela M. J., Papalambros P. Y. and Ulsoy A. G. (1985). Programming optimal suggestions in the design concept phase: application to the Boothroyd assembly charts. *Journal of Mechanism, Transmissions and Automation in Design*, **107**

Moinet M. (1986) New dimensions in automation. In *Proceedings of the 7th ICAA*, Zurich, IFS (Publications) Ltd

Poli C. and Graves R. (1985). *Assembly Analysis and Line Balancing Spreadsheet*. Amherst: University of Massachusetts

Poli C. and Fenoglio F. (1986) *Automatic Assembly Spreadsheet*. Amherst: University of Massachusetts

Poli C. and Groppetti R. (1986) Economic justification for automatic assembly based on the capital cost of equipment. In *Proceedings of the 7th ICAA*, Zurich. IFS (Publications) Ltd

Poli C., Graves R. and Groppetti R. (1986). Rating products for ease of assembly. *Machine Design*, **21**

Ramsli E. and Neerland H. (1986). Economic modelling of flexible assembly systems. In *Proceedings of the 7th ICAA*, Zurich, IFS (Publications) Ltd

Swift K. G. and Firth P. A. (1984). Knowledge based expert systems in design for automatic handling. In *Proc. of the 5th ICAA*, Paris, IFS (Publications) Ltd

Takahashi K. and Senba K. (1986). Design for automatic assembly. In *Proc. of the 7th ICAA*, Zurich, IFS (Publications) Ltd

Withney D. E., Nevins J. L. and Graves S. C. (1984). *Applying Robots in Industrial Assembly*. Report, MIT Industrial Liaison Program, Cambridge, MA

Acknowledgements

This work is part of a research supported by the Italian Ministry of Education. The authors are grateful also to DEA – Digital Electronic Automation, Moncalieri (Torino), Italy, for having kindly provided industrial study cases.

4.5

Towards integrated design and manufacturing

M. Susan Bloor,
A. de Pennington,
S. B. Harris,
D. Holdsworth,
Alison McKay and
N. K. Shaw
University of Leeds, UK

■ Introduction

Currently much is written of attempts to integrate pockets of com-puter-aided activity because there is a consensus that the real benefits to an enterprise of automation technology will only then be realized. This integration implies integration of the organization's information and information flow. Research into aspects of CIM, and architectures to support it, is being undertaken by, for example, ESPRIT, Alvey and US Air Force initiatives, CIM OSA (1987a and 1987b), ANSA and EIS (1987a, 1987b and 1987c), respectively.

In CIM a fundamental requirement will be a model to support the life cycle of the product. This product model is a superset of the models used in current CAE systems which are centred around product data, loosely that data represented in engineering drawings. The work reported here reflects the authors' concerns for the information content, the what, rather than the distribution mechanisms, the how, of a CIM architecture.

This paper is reprinted by permission of the authors and the University of Leeds from *Factory 2000*, Cambridge, Sept. 1988, pp. 21–9

The essence of product definition data is that it is highly structured. For instance, an assembly is a hierarchy of subassemblies and components and a network may be formed by components which are common to more than one subassembly: a car may be made up of a chassis, an engine, a gear box and so on while the same nut may be used in the gear box and the engine. In order to manipulate such data, the user (engineer) needs structural awareness as well as spatial awareness. The latter is one of the hallmarks of a good engineer and allows him to manipulate geometry. Similarly, people seem to have widely varying capability for handling structural information. The core of the PDS is a structure editor which provides an interactive environment to complement human structural insight. The need to allow a view of the data which is not directly related to its physical representation in the computer is discussed in Hull and King (1987). The structure is the primary data: interactive users work with the structure itself rather than some representation which is based upon it. A fuller description of the structure editor, SE, is given by Holdsworth (1988). However, to enable the reader to understand the experiments, a brief description is given in a later section.

Three of the experiments which have been conducted with the software are presented:

- a first experiment in integration;
- the development of product data for a gudgeon pin, which illustrates the descriptive and parametric capabilities of the SE;
- the definition of part of the STEP (ISO, 1988d) product model, which shows the relative rôles of a product model and an editor designed to allow the manipulation of structured data.

■ Product modelling

□ The form of the data, the product model

The term 'product model' first gained acceptance in the context of international efforts to develop a neutral format (for the transfer of product data between computer-aided systems) which is capable of representing product data for all sectors of engineering as well as all aspects of the life cycle (IGES, 1986). Some of the main features of the data are described in detail in Madison *et al.* (1988) and possible ways in which such types of data may be represented in a computer are discussed in Atkinson and Buneman (1987). The place of a product model in a CIM environment similar to that of the authors is described in Iwata and Sugimura (1986).

Geometry is one of the factors which contributes to the complexity of processing engineering data compared with that of most other areas. Other factors are:

- the interdependence of the data, constraint management is recognized as a major problem;
- the need for parametrized representations to facilitate the definition of new products;
- the evolutionary and iterative nature of design; and the subsequent need to maintain potentially inconsistent data until the end of the design process when a single consistent design has been achieved;
- the numerous versions of products which result from engineering change requests and marketing a variety of products;
- the need to explore the implications of change.

It is now accepted that a complete representation of geometry, which is a basic physical attribute of engineering components, is necessary for many applications. The most appropriate representation and the need for multiple consistent representations within one system to ensure that geometric functionality can support applications are discussed elsewhere (Kane, 1985). The geometric underpinning is the most developed part of the ISO/STEP product model (ISO, 1988d, Section 1): the form and content of the full product model is still a research area to which the authors' work contributes.

The form of the data within the product model falls into two categories: one category will be used by software and the other by people. Software must be aware of the form of the data which it will use, so the structure of this data must be fixed in the product model. The data which people use is typically at a higher level of abstraction and its structure may be changed to suit the needs of individual people or companies: such data structures will usually be defined in terms of the fixed data structures. For example, the geometry data structure may be defined in terms of the Boolean operators and half-spaces which are passed to the geometry applications. People would define geometry in terms of more natural entities, such as blocks, cylinders, bosses and pockets.

Each class of data may be either product specific, company specific or reference data. Product-specific data is that data which applies to a single product, for example, its fully defined geometry. Company-specific data is that data which applies to more than one product, for example, the manufacturing facilities will probably be used to manufacture several products. Reference data is more general, for example, material properties and data which is held in standards. The means of defining and manipulating the product model should be flexible enough to allow new forms of data and applications to be added in the future.

☐   Data manipulation, the structure editor

The SE is a generalized tool for defining data structures and building interfaces between those data structures and application programs which have been written independently of the data structures and the SE. The basic form of the data, generally that used by software, is described by a meta-structure. A well-used meta-structure is one which describes assemblies, the structure of which is shown in Figure 1. Instances of this meta-structure define actual assemblies. The meta-structure could be extended to describe the data needed by other applications and existing instances could still be used. For example, if a company had a paint shop then the component may need an extra field to specify its colour: the colour could then be used to determine paint requirements.

The logical relationships inherent in the data, for example, that an assembly has subassemblies and components and that components have geometry and material, are captured and presented to the interactive user. This gives the user the impression that he can 'touch' the data structure (Powell, 1987) and allows both the structure and content of definitions to be visualized.

Data sharing allows the relationships between different parts of the description to be defined and then maintained automatically. For example, the nominal diameter of a nut can be shared with that of the bolt on to which it will be fitted so that if its value is edited the change will be reflected in both places. Data values are forced to conform to the expected types during the editing process: this is the strong typing whose value is recognized in programming languages such as Pascal and Ada. It ensures that only correctly formed data can be defined and passed to applications. Such a

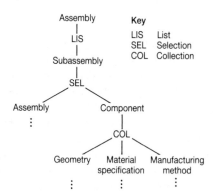

Figure 1   A meta-structure to describe assemblies.

facility can also be found in current language-sensitive editors such as VAX LSE (DEC, 1987).

The SE provides a parametric capability which allows interactive users (engineers) to define product data in terms of the higher-level entities described earlier. Any piece of data structure may be parametrized and the parameters may be of any type, simple (integers, reals and strings) or structured (lists, choices and records). This contrasts with many of today's CAD systems which offer limited parametric capabilities, such as, only numeric parameters. For example, a component may be defined in terms of its geometry, material and cost; a mild steel component would have the cost and geometry as parameters; a turned mild steel component would have cost and its profile as parameters and so on. A powerful parametric capability enables both the evolution of a design and the use of existing ones, or parts of them.

Data may be made available to a set of applications programs, including those written by third parties providing their data requirements are sufficiently understood. The SE produces Ada code which can be used to extract data from and, to a lesser extent, build data structures. The generated code includes an Ada package specification for the interface to the application routines. Interfacing is achieved by writing the body of this package, much of which is straightforward: in this way interfaces to software which use data described by the product model can be built. The specification and body mechanism (as in an Ada package) provides a good software integration mechanism.

■ Case studies

□ Integration

This experiment demonstrates, using the description of a 2½ D machined component, that the data required to support the manufacture of such a component can be held within a single product description. The basic data structure shown in Figure 1 defined the form of the data. A component was described using the SE and graphical representations of all or a part of it were produced by applications which used the geometric data. An interface to a third party proprietary database gave access to tooling data which simulated a company specific database. The NC code needed to manufacture the component was generated by manufacturing applications, cutting data selection, cutter path generation and automatic machine tool set-up, which used the component description and database data. Figure 2 shows a subset of the system which was used to test the ideas. The boxes show some of the geometry and manufacturing applications which were written by the project team while items in circles are external to the system and would vary depending upon the company in which the system was implemented. Lines

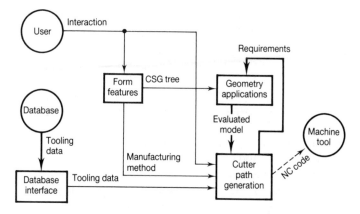

Figure 2   The integration experiment.

represent data flow: the solid lines show flow through the SE, dashed lines indicate that other means were used, for example, a database query language or a direct numerical control, DNC, link, and dotted lines show direct links between software packages.

The SE was used to bring together several discrete pieces of software. All the applications used the same data model. A database interface added what could have been company-specific data to that model. The Ada code-generation facilities were used to build the interfaces between the applications and the data, including the database interface: other interfaces could be added in the same way. The demonstration was limited to 2½ D geometry due to the functionality of the manufacturing applications which were available at the time. Future experiments will explore the problems which emerge when more complex 3D geometry is handled. It is recognized that more flexible interfaces to other systems will be needed in the future, for example, the ability to update databases and have real-time links between systems.

The experiment was carried out by a team of 10 software engineers and, as a result, the system we produced is a melting-pot of styles and ideas. With hindsight we can say that the software tools, such as the SE and high-level programming languages, which we used are not all that is needed to support the building of large, multi-author software systems. Methodologies for the use of these tools are also needed if a more co-ordinated approach is to be taken.

☐   **Product specification**

The description of a typical component contains data such as its geometry, raw material, manufacturing method and cost. A gudgeon pin is such a component: all gudgeon pins have data which is common to their class, for

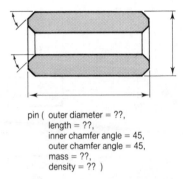

pin (outer diameter = ??,
      length = ??,
      inner chamfer angle = 45,
      outer chamfer angle = 45,
      mass = ??,
      density = ??)

Figure 3 The specification of a gudgeon pin.

example, they are all tubular and manufactured in similar ways. Since the specification of all pins is the same and the relationship between the specification and the body of the definition is known, all pins may be defined in terms of a set of parameters. Other data may be derived from these parameters either as the results of calculations or by looking up values from tables of data. Figure 3 shows the geometric parameters which were used to specify a gudgeon pin. Parametrized pieces of structure are implemented in the SE as functions. When invoked, a frame for the pin specification of the form shown in Figure 3 is presented to the user. It can be seen that the chamfer angles have default values and that all the other values are undefined. The user must assign values to the undefined ones and may change the defaults.

The body of the SE function, the parametrized definition of the gudgeon pin, is defined in terms of the parameters which are normally used to specify a pin. For example, the mass, density and specified dimensions are used to calculate the wall thickness and the dimensions of the tooling used in its manufacture because a customer specifying a pin gives the length, outside diameter and maximum allowable weight rather than the wall thickness. Look-up tables are also implemented within the body of the function. For example, the value of the wall thickness is used as a key to tables which define the tolerance on the wall thickness and the length of the internal chamfers. Figure 4 shows the nature of the links between the formal parameters of the pin, its definition and the tables of data. Solid lines show links between the body of the definition and the parameters. The tables are tied to the definition in two ways: dashed lines show parts of the definition which are used as keys to tables and dotted ones indicate parts of the definition which are defined by data from tables.

This example has demonstrated the flexible parametrization facilities of the SE: any piece of definition may be parametrized, interactively and with any number and types of parameters. Strong typing ensured that all the pins defined had the correct number and types of parameters. It can be seen that all the parameters were numeric but they could have been any type of structure,

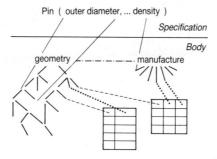

Figure 4   The structure of the pin definition.

as could the result type of the function itself. For example, in the future the density could be replaced by a material specification. When assigning a value to the material specification, the user could be offered a menu of standard materials to pick from and the result of his choice would allow the density to be retrieved from a material's database. The SE allowed look-up tables to be modelled. In the future, such data could be stored in a database and referenced, rather than a copy being stored with the body of the pin definition.

☐   Rapid prototype testing of standards' schemata

The activities of the ISO committee TC184/SC4/WG1 are aimed at producing a new standard for product data exchange (Standard for the Exchange of Product Model data – STEP (ISO, 1988a). The main output of the committee are conceptual schemata for various aspects of product data (ISO, 1988d) defined in a data description language called Express (ISO, 1988c). A significant problem faced by TC184/SC4/WG1 is that of determining the quality of the schemata. Currently, the principal available test is to populate an implementation form and perform queries on the resulting database. However, this involves considerable effort in three areas:

- obtaining a clean mapping into the database description language from the schema in Express;
- entering the data;
- defining an appropriate user interface plus relevant queries.

The SE has been found to provide an alternative software basis for a trial of a given schema:

- The mapping between Express entity definitions and the SE's meta-structure is straightforward. For example, the Express entity product item is described by its name, description, part number and versions and maps directly on to an SE collection of the attributes.

- As each item of data is entered there is immediate feedback to the user in terms of completeness, typing and suitability of names (semantics).

- There is no call to write interface code. Users can navigate the resulting instance structures and formulate queries as appropriate.

A view of the data which resembles closely the original Express definition can be defined.

Here we have explained how the SE can provide a rapid implementation path for, and enables the schema designer to assess various characteristics of, the proposed structure. This is achieved by defining a meta–structure which comprises a list of data sets. Each data set is a collection of subschemata where each subschema is a collection of lists of individual entities. Cross-references between Express entities are resolved by sharing in the meta-structure. Sharing of data at the instance level in the SE corresponds closely to concepts of internal expansion (not shared) and external reference (shared) applied to entity attributes in Express and in the proposed physical file format for STEP (ISO, 1988b). With future developments, it should be feasible to automate the translation from schemata in Express to the SE's meta-structure. The SE's code-generation facility can then be used to facilitate the generation of STEP physical files with known content without using a particular CAD system.

■ Discussion

Our work has concentrated on the data modelling rather than system architectural aspects of CIM. The ANSA project, for example, is aiming to formalize the interface between applications on a distributed system to reduce dependence on system elements such as programming language and operating system. At the highest level, an interface definition language, IDL, in which the functionality of the interface and the data to be passed may be expressed is defined. Currently this data is limited to simple types such as integers, reals and strings. The Alvey design to product demonstrator project (Burrow, 1986) has a similar architecture for connecting application packages, with a tool manager controlling the connections between packages. This also requires an IDL, which is currently being defined, as well as a product description and an engineering knowledge base: the IDL must be able to represent the structured data in the description and knowledge base. As with the MAP specification (General Motors, 1987) for shop-floor communications the structural relationships between these simple data representations is left to the applications. The data–modelling view is that the relationships between these simple types is what captures the design intent. The authors have presented their experiences with a tool which was designed to aid the creation and editing of structured representation of product data.

We have described a general-purpose tool to help investigate the form and content of data. It has already proved useful in the Leeds CAE group's contribution to the product model which is being evolved by ISO and represented in a formal language, Express. It is encouraging how closely the constructs of Express parallel those we have found necessary in our modelling to support mechanical engineering design and manufacture. Several areas, including the in-flight checking of constraints, have yet to be addressed. Work on enhancing the SE to provide graphical presentation of the structures is underway. Other systems appear to be graphically superior but lack the interactive binding and strong typing which we believe are so valuable.

The relationship of this work to the object-oriented paradigm is being investigated. Data and rudimentary methods may be represented using the functional capability of the SE. Because the software was designed to explore structured data, any form of data structure can be incorporated by specifying the form in a bootstrap meta-structure, the start-up point, which currently has lists, collections, selections and simple types. In particular, facts and rules which represent knowledge have been described in terms of the bootstrap meta-structure; presently, however, there is no processing software.

■ The future

It seems unlikely that anyone can predict exactly what form computer systems in engineering will take in the year 2000. For the purposes of our discussion we will use the acronym CAA to stand for CIM applications architecture. (cf. SAA, IBM's systems applications architecture which was announced as the common interface for applications software right across the IBM range. At present, very few details about this interface are published.)

The long term view of our work is summed up in Figure 5. The shaded central box represents the architecture into which closely coupled applications are connected (represented by rectangles). These applications will be written with a knowledge of the nature of this CIM architecture, and will be able to exploit its capability. For at least a decade there will be significant use of software systems which are implemented outside CAA. Such systems (shown as circles in Figure 5) will need some customized interface by which they are loosely coupled on to CAA if they are to be used in the CIM environment. The capability of interfaces will vary widely. The simplest interface would be capable only of recording in the CAA data the nature of the information stored in the loosely coupled application. More complete interfaces would offer the capability for translating the product data between

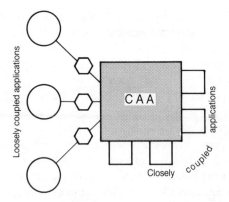

Figure 5 CIM applications architecture.

its CAA form and its application-dependent form. The CAA data will contain all the persistent information used by the closely coupled applications and it is conceivable that the ISO product model may become the standard schema in a CAA.

■ Conclusion

We have concentrated on the data-oriented aspects of CIM, believing that the central issue is the provision of an environment in which one can build models of products. A prototype system to help engineers to manipulate product data has been created. This prototype explores the use of computational structures and their interfaces to the end user. In the future the relationship of our work with emerging techniques, such as object orientation and knowledge-aided systems, will be better defined. In a similar vein, an architecture for CIM will be adopted with the objective both of housing new tools and interfacing to existing ones.

References

ANSA. *Advanced Network Systems Architecture (ANSA) Reference Manual*. Alvey Project No. LD007

Atkinson M. P. and Buneman O. P. (1987). Types and persistence in database programming languages. *ACM Computing Surveys*, **19**(2), 105–90

Burrow L. D. (1986). The design to product alvey demonstrator. *Electrotechnology*

CIM OSA (1987a). *A Primer on Key Concepts and Purpose*. ESPRIT Project No. 688

CIM OSA (1987b). *Strategic Management and Design Issues.* ESPRIT Project No. 688

DEC (1987). *Guide to VAX Language-Sensitive Editor and VAX Source Code Analyzer.* Digital Equipment Corporation

EIS (1987a). *Interface Requirements' Specification.* Honeywell Systems and Research Centre, Minnesota

EIS (1987b). *Software Requirements' Specification.* Honeywell Systems and Research Center, Minnesota

EIS (1987c). *System/Segment Specification.* Honeywell Systems and Research Center, Minnesota

General Motors (1987). *MAP V3.0 Specification.* General Motors, implementation release July 1987, final release due September 1988

Holdsworth D. (1988). Lambda capability in a structure editor. *Under preparation*

Hull R. and King R. (1987). Semantic database modeling: survey, applications and research issues. *ACM Computing Surveys*, **19**(3) 201–60

IGES (1986). *PDES Initiation Activities.* IGES Committee

ISO (1988a). *Preliminary Design Document.* ISO TC184/SC4/WG1, document N208

ISO (1988b). *The STEP File Structure.* ISO TC184/SC4/WG1, document N219

ISO (1988c). *Information Modeling Language Express.* ISO TC184/SC4/WG1, document N224

ISO (1988d). *STEP/PDES Conceptual Schema Testing Draft.* ISO TC184/SC4/WG1, documents N241, 242, 243

Iwata K. and Sugimura N. (1986). Development of product model for design and manufacturing of machine products. In *NAMRC XIV proceedings, SME*, 528–34

Kane P. M. (1985). *Provisional Report on a Geometry Database for the Design to Product Project – Part 1: A Review of Previous Work.* Technical Report, University of Leeds

Madison D. E., Wilbur T. G. Jr and Thomas Wu C. (1988). Data-driven CIM. *Computers in Mechanical Engineering*, **6**(5) 38–45

Powell M. S. (1987). Strongly typed user interfaces in an abstract data store. *Software – practice and experience*, **17**(4) 241–66

Acknowledgements

The authors wish to thank their many colleagues in the CAE group at the University of Leeds, the Department of Manufacturing Engineering at Loughborough University of Technology and members of the sponsoring companies. Our understanding of what constitutes a product model has benefited from close co-operation with those researching application areas.

This work has been funded by DTI Agreement Nos. RD142/08, MEE/68/06 and MMT/68/07 since 1979 and the ACME Directorate of SERC Grant No. E04301 since 1987. Discussions with the team on the Alvey design to product demonstrator project have been mutually beneficial.

The gudgeon pin study was the subject of a final year linked project with Hepworth and Grandage Limited (Bradford), executed by Jonathan Stancliffe and supported by their staff. We acknowledge their efforts.

Part 5

Strategies and Organization

■ Introduction

Hayes *et al.* (1988) rebuke certain kinds of engineers whom they call 'handbook' engineers. In the hands of such people, DFM would simply become something you look up in a book – in a book even such as the present one! 'Find out how to do design for manufacture, then put it into practice in our firm', the managing director might be heard to declare. The so-called 'handbook' engineer goes to it with a will, and the result is something to do with DFM, but not necessarily something that helps the firm.

The final part of this book will indeed offer some detailed suggestions of a 'handbook' nature, but its main thrust will be to highlight broad ways of running manufacturing businesses so that they become more and not less competitive. In particular, this part sets out to support a thesis that design for manufacture has to be adopted *consciously* within the organization, that there has to be a *balanced* approach to product and process development, and that manufacturing firms actually have to change.

We will start by looking at some solutions to engineering problems achieved by a new company and then look at the organizational means that the company thought were appropriate.

■ Discussion of papers

The article by Stevens and Tawil describes how Lister–Petter Limited deliberately designed for manufacture a new range of industrial diesel engines. The case study shows typical DFM principles and techniques at work achieving, for example:

- a reduction in the number of machining operations required on the crankcase;
- common usage of intake and exhaust manifolds in the cylinder head;
- improved design of the valve and gear train for assembly;
- a reduction from several types of pumps for various models to one common pump for all models.

Figure 1 in the case study shows a flow diagram of how the new range of engines was launched two-and-a-half years after the production of a design concept. The first two sections of the case study describe how the company approached its task.

In order to assess the methods adopted by this company, we need to have some kind of yardstick for comparison purposes. Such a yardstick is available in the work of Hayes *et al.* (1988). Basically, these authors try to address the question that asks why companies do not transform their manufacturing function so as to gain competitive advantage. Their research identified what they call the 'conventional' and 'new' paradigms for product and process development projects. Companies using the conventional paradigm may lose product leadership and market share, while companies using the new paradigm are effective at exploiting technology and shortening lead times for bringing new products to market. Table 1 contrasts conventional, CP, and new, NP, paradigms according to Hayes *et al.*

Lister–Petter Ltd certainly seemed to be good at doing DFM, but a further reading of the case study reveals that it was only working to the new paradigm in part, and that in some respects the procedure owed more to the conventional paradigm. Our commentary now looks at what Lister–Petter did, to ascertain in a broad-brush fashion how it operated its product and process development project. To some extent, the priorities identified by Lister–Petter's market survey were non-negotiable (CP). The company was strong on stakes such as project cost and product performance specification (NP), but other stakes, such as process performance specification and project schedule, only entered later (CP). It is not clear from the case study who led the cost control management team, but the team does seem to have been genuinely cross-functional (NP). Project phases tended to be sequential (CP) but there was sufficient overlapping (NP) for product design to be modified early in the proceedings.

What emerges from this analysis is that Lister–Petter was not working entirely according to the new paradigm (NP) as far as organization and strategy were concerned. It knew plenty about DFM, but judged by Table 1 at any rate it was still working to a conventional paradigm in some respects. That's a very broad statement, and one that the company might wish to challenge. In order to achieve certain DFM goals (which were undoubtedly achieved), various things had to happen. One of these was integration of personnel, and this is one of the most important tasks that Lister–Petter must have achieved successfully.

The case study states that 'on this project the various experts worked full-time being totally involved with the designers at the CAD screen to ensure that every aspect of the design was optimized'. This sounds like some sort of working group or 'cell', and the next paper by Fairhead is specifically about creating a cell.

Fairhead approaches the creation of a cell strictly from the point of view of the human beings involved. For large successful companies, such as Ford, to create a cell may be relatively easy, but in unsuccessful firms it is difficult. Two of the needs that arise are for a product sponsor to fight political battles on behalf of the cell and a product champion to lead the work itself. As far as the DFM cell is specifically concerned, a charter for a multi-functional team was proposed in Part 2 by Whitney. Whitney's charter is one way of doing it and the need for departmental walls to be bridged has been discussed.

Fairhead's snappy and anecdotal approach can really be applied to any

Table 1 Contrasting the conventional and new paradigms for product and process development.

Dimension	Conventional paradigm	New paradigm
Reference stakes	Non-negotiable, set early	May require occasional revision
Project cost Product/process performance specs Project schedule }	One seen as primary driver for the firm	All given significant attention
Project team	Led by design engineering	Led by broad, experienced business manager
Project focus	Starts in marketing, shifts to engineering, ends in manufacturing	Cross-functional team effort throughout
Project phases	Sequential	Extensive overlapping
Management obstacles	People transferred; priorities changed; bottlenecks affect limited design resources; slow turnaround; vendor delays; excessive engineering changes	Managed through better plans, discipline; skills and follow-through
Responses to schedule slippages	Deny slippage has occurred; skip steps; announce completion, hand-off to operating organization	Much less frequent, but addressed realistically as they arise
Handling of key tasks		
Problem solving	Within functional group or discipline	Overlapping and cross-functional
Conflict resolution	Suppressed, postponed or sent upstairs	Addressed early and at low levels
Project organization	Primarily functional with hand-offs	Heavyweight project managers maintain integration
Manufacturing control	Only lower forms of control considered necessary	Progressive or dynamic control necessary
Information transfers	Large batches, transferred downstream only after completion of phase	Many smaller two-way exchanges throughout development
Early manufacturing involvement	Looked upon as (undesirable) constraint	Based on trust and mutual respect; adds value
Design for manufacturability	Gives manufacturing veto power over design engineering	Provides improved products and processes

Reprinted by permission of The Free Press, a Division of Macmillan, Inc. from *Dynamic Manufacturing: Creating the Learning Organization* by Robert H. Hayes, Steven C. Wheelwright and Kim B. Clark. Copyright © 1988 by The Free Press.

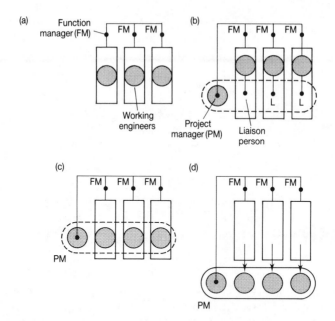

Figure 1 Types of organizations for development projects: (a) functional organization; (b) lightweight project manager; (c) heavyweight project manager; (d) tiger
team organization. Reprinted by permission of The Free Press, a Division of
Macmillan, Inc. from *Dynamic Manufacturing: Creating the Learning Organization*
by Robert H. Hayes, Steven C. Wheelwright and Kim B. Clark. Copyright © 1988 by
The Free Press.

team or cell situation – it is not specific to a DFM cell. His Venn diagram way
of displaying the dynamics of the cell is an ideal, typical illustration of just one
aspect – namely the boundaries of expertise. We must look at another set of
diagrams (Figure 1) from Hayes *et al.* to comprehend the range of team or cell
structures, and to consider which type of team/cell is most likely to succeed.

According to Fairhead's way of looking at things, Figure 1(a) would
be least likely to succeed. There is no integration of functions, and there are
demotivating 'walls' between functions. Similarly, with (b) there is no
integration, but at best a kind of liaison committee overseeing things.
Something like Fairhead's cell is emerging in (c), and a cell or team does
emerge in (d). To achieve the ideal, typical interaction of expertises within
the cell, then the circles in (d) should overlap, as indicated in Figure 2.

Hayes *et al.* are in no doubt as to which of the team structures (a–d) are
best for new development. They say (p. 321): 'Although each of these
organizational forms has strengths and weaknesses, our research suggests
that the heavyweight project manager and the tiger team are the most
effective.'

The Fairhead extract is valuable because it presents a series of tips on

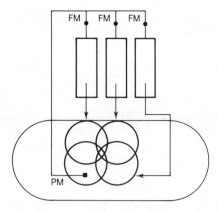

Figure 2   Overlapping expertise within a 'tiger team' structure.

how to integrate experts. Such integration was important to Lister–Petter when it attempted DFM, and such a structure as Figure 2 is more likely to be effective according to the research of Hayes *et al.*, taken together with the Fairhead extract.

Creating the multi-functional team is an essential part of the new manufacturing paradigm, because processes as well as products need designing. Only by bringing experts together in this fashion can process and product development interact constructively.

If we look again at Table 1, it is immediately apparent that the diagram is contrasting paradigms not only for *products* but also for *processes*. There is an underlying tension between product and process development involving trade-offs, which needs to be explicitly recognized. DFM can only be successful if rational trade-offs are planned and actually take place.

The next paper by Hayes and Wheelwright is a classic article on products and processes. The authors are of course the same authors from whose later work (Hayes *et al.*) we have already quoted. The 1979 article is an example of seminal work from the Harvard Business School which laid the foundations for better approaches to manufacturing strategy. These ideas are continuously emphasized by Skinner (1969), Skinner (1974), and Hayes and Wheelwright (1984). In the UK, the ideas have been taken up and developed in many manufacturing firms by Hill (1985). The most recent studies in concurrent design (for example, Nevins *et al.*, 1989) also draw on this key work.

The basic message of this particular article from the Harvard Business School is that the interaction between product and process is crucial. Figure 2 in the article sets out to show traditional manufacturing modes plotted against the life cycle of products. Subsequent work by these authors and others devised rigorous techniques for analysing product and process needs together. That is, in fact, what Lister–Petter did up to a point when it designed its diesel engines for manufacture.

When looking at Table 1, we notice that DFM merits just one line only – at the bottom of the table. According to these authors, the conventional way of doing things was to give manufacturing the veto power over design engineering. The new paradigm, they claim, provides improved products and processes. To sum up then, the Harvard Business School work is important to the practice of DFM because it puts DFM firmly where it belongs – as part of the tension between product and process development – a tension that has to be recognized and dealt with.

With the significance of product/process tension well in mind, we now return to a further examination of how to integrate people within the company. In their introduction, Ettlie and Reifeis refer to problems at the design–manufacturing interface – this, we think, is just another way of describing product–process tension. The authors quote from Stoll's advice, which we already looked at in Part 3. They then provide a series of small case studies indicating the nature of the DFM being attempted and the changes in the organization that DFM required. The storylines are familiar, the means to success perhaps less so. Table 1 in the article distils five methods used by these companies to promote integration, as follows:

- Design manufacturing team.
- Compatible CAD systems for design and tooling.
- Common reporting position for computerization.
- Philosophical shift to design for manufacturing.
- Engineering generalists.

A philosophical shift to design for manufacturing requires, in the words of these authors, 'a significant educational' program. The second paper from Fairhead presented in this part gives a further set of tips for achieving such a program, only he does not call the program 'educational', he talks of using strategy to create culture – *new* culture. Forces against change tend to be strong within manufacturing industry. Trying to shift to DFM may be met by misbelief arising from unchallenged folklore. But 'an effective design and innovation culture is, by definition, open and responsive'.

Fairhead goes on to explore several ways in which the *status quo* may be challenged. DFM will never get off the ground unless old ways of doing things are challenged. Getting specialists to overlap their competencies in the NPD cell has already been referred to in the first Fairhead paper. To sum up what Fairhead is getting at in the second paper, we can say that he recommends structured interventions from the top down so as to break belief systems that do not think design for manufacture is either a good thing or possible.

The final paper by Dean and Susman neatly summarizes and clearly restates with examples some of the important themes and axioms explored earlier in this book and in this part, such as:

- Final designs emerging from engineering may be producible only at a very high cost.
- Effective manufacturers aim for fewer and more standard parts.
- DFM cannot be achieved by exhortation.
- Encourage interaction between specialists at an early stage.
- Take care about using a new type of specialist (for example, the 'integrator' or 'liaison' specialist).
- Set up cross-functional teams and new product development cells.
- Consciously determine the trade-offs between product and process development.
- Practise simultaneous engineering.

Most effective action on DFM will fall within the eight themes/axioms listed here.

Finally, this part has supported the thesis that design for manufacture has to be adopted *consciously* within the organization. There has to be a properly *balanced* approach to product and process development. Firms themselves have to change if any progress in this direction is to be achieved.

References

Hayes R. H. and Wheelwright S. C. (1979). The Dynamics of process product life cycles. *Harvard Business Review*, March–April

Hayes R. H. and Wheelwright S. C. (1984). *Restoring our Competitive Edge*. New York: John Wiley

Hayes R. H., Wheelwright S. C. and Clark K. B. (1988). *Dynamic Manufacturing: Creating the Learning Organization*. New York: The Free Press

Hill T. (1985). *Manufacturing Strategy: The Strategic Management of the Manufacturing Function*. London: Macmillan

Nevins J. L. *et al.* (1989). *Concurrent Design of Products and Processes*. New York: McGraw-Hill

Skinner W. (1969). Manufacturing – missing link in corporate strategy. *Harvard Business Review*, May–June

Skinner W. (1974). The focused factory. *Harvard Business Review*, May–June

5.1

A new range of industrial diesel engines with emphasis on cost-reduction techniques

J. L. Stevens and
N. Tawil
Lister–Petter Limited, UK

■ Introduction

To establish the future worldwide requirements for small diesel engines, in the power range from 8 to 30 kW, a massive independent market survey was commissioned by Lister–Petter and completed in 1985. The outcome of this research programme was the clear identification of a need for a totally new range of engines, incorporating specific design features to satisfy the future demands for industrial, agricultural, marine and construction industry applications.

A major design study was therefore carried out at Lister–Petter with a view to satisfying the basic requirements identified from the market survey, which had highlighted the following priorities from a wide variety of users and original equipment manufacturers:

(1)   low cost;

(2)   low noise and vibration;

(3)   high level of durability and reliability;

(4)   ease of maintenance;

This paper is reprinted by permission of Lister–Petter Ltd from the paper written by the authors, 1986, pp. 1–12

(5)   compact size and low weight;

(6)   air- and water-cooled options;

(7)   high degree of common components.

■ Total company involvement

To ensure that these requirements were totally satisfied it was considered essential to change previous practices within the company in order to meet the low-cost objectives and also to reduce the time period from design concept through to production. To this end a cost control management team was established, consisting of experts from production, foundry, sales, purchasing, marketing and finance, to work full-time on the project in conjunction with the design team, to ensure that every aspect of the design was carefully studied and 'designed for manufacture'.

It is normal to involve experts from the various company disciplines at the design stage on an 'as and when required basis' but on this project the various experts worked full-time, being totally involved with the designers at the CAD screen to ensure that every aspect of the design was optimized. This continual optimization process ensured that the number of machining processes was minimized. Every machined face, for example, was investigated to explore whether it could be eliminated or produced in a different way to reduce the time taken and cost to produce it. As a result very significant cost reductions were achieved, resulting in some component costs being less than 50% of previous designs.

At Lister–Petter we are fortunate in having our own iron foundry, and the direct involvement of the foundry expert at the design stage proved to be a considerable advantage, as it allowed complete optimization between the machining processes and casting feasibility for lowest cost.

Inevitably, this direct involvement by the foundry necessitated new technologies which were successfully incorporated and consequently improved its competitiveness, as a supplier of castings, both within the company and to other engine manufacturers.

■ Company plan

Having established the necessity for total company involvement to ensure that the objectives could be addressed appropriately, a project plan was prepared by the cost control management team which identified the key decision dates, as shown in Figure 1.

This master plan was referred to continuously throughout the project

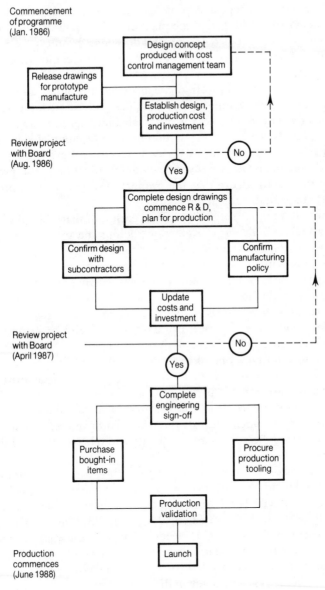

Commencement
of programme
(Jan. 1986)

Design concept
produced with cost
control management team

Release drawings
for prototype
manufacture

Establish design,
production cost
and investment

Review project
with Board
(Aug. 1986)

No

Yes

Complete design drawings
commence R & D,
plan for production

Confirm design
with
subcontractors

Confirm
manufacturing
policy

Update
costs and
investment

Review project
with Board
(April 1987)

No

Yes

Complete
engineering
sign-off

Purchase
bought-in
items

Procure
production
tooling

Production
validation

Production
commences
(June 1988)

Launch

Figure 1

to enable the various departments throughout the company to prepare and
organize their activities to satisfy the required objectives. It was also reviewed
weekly by the committee to ensure that the project remained on course, and
that any necessary corrective action was taken.

The company plan adopted, consisted of three essential phases: design
concept and proving, engineering sign-off and production validation.

The design concept and proving phase was to determine that the concept was viable and satisfied the main objectives laid down. It involved basic design layouts, detailed designs and drawings, prototype manufacture and initial performance tests to confirm that the base engine design criteria were satisfactory.

The engineering sign-off phase included the build and test of 20 engines of each type. Some of these engines were used for performance development and the remainder for either reliability development, durability or early field trials. The total engine running hours accumulated in this phase was approximately 20 000 hours for each type of engine. This phase also confirmed the manufacturing policy and introduction of a production feasibility system to ensure that the production department approved the design as being suitable for manufacture.

The production validation phase ensured that the product was satisfactory when built with production tooling and production assembly techniques. Here again 20 engines were built and tested to a variety of test and endurance programmes where over 20 000 hours of testing were accumulated on each type in order to confirm the performance and reliability of the product from production.

During the three phases any fault found with the engines in respect of performance and reliability, and even any item found to be aesthetically offending, was raised by a formal reporting document. Any person within the company could raise one of these documents and the product was not given final release until each fault or comment was satisfactorily cleared through the design manager back to the originator.

This tight control method ensured that the final product was satisfactory before launch and that the whole company was satisfied that it would reach the highest standards of quality and reliability demanded in the market place.

As a result of this strategy the company remained totally committed to the project and many new practices were introduced. For example, it became immediately apparent that there would be a bottleneck in the design office if the time-scales were to be satisfied, despite the high investment costs in providing a comprehensive CAD system. The decision was therefore taken with full support from the staff to introduce a two-shift system within the office to provide design capability for 16 hours a day.

Similar changes were made in the production areas, and as the engine test facilities were seen as insufficient to meet the objectives, new facilities were introduced both in the research and development, and production areas to allow for fully computer-controlled, 24 hours a day, engine testing, with performance monitoring and automatic engine test programme control.

This basic programme generated the company attitude towards satisfying the objective of introducing the new product range in the shortest time, which also reduced total company costs.

■ Design philosophy

Taking into consideration that the power objective for the new engine range was 8–30 kW for continuous speeds up to 3600 rpm, for both air and water cooling, the first essential decision to be made was the size and number of cylinders to cover the power range. After much deliberation it was decided to commonize on one stroke of 80 mm throughout the range and provide two-, three- and four-cylinder versions for both air and water cooling, bearing in mind that the maximum number of components in both types of engines needed to be common, in order to reduce cost as well as to satisfy one of the objectives.

The next key decision was the distance between cylinder centres on each of the engines and again, after careful consideration, 100 mm was used on both air- and water-cooled engines to allow for common machining facilities. With this cylinder centre dimension, an 86 mm cylinder bore was feasible for the water-cooled engines and 76 mm cylinder bore for the air-cooled engines. The reduced bore size for the air-cooled engines was due to the necessity of providing cooling fins on the external cylinder diameter. This resulting geometry gave a bore stroke ratio of 1.07 for the water-cooled and 0.95 for the air-cooled engines; these were considered to be an acceptable compromise, keeping common the geometry for the lower part of the engine. With these cylinder dimensions and the known achievable brake mean effective pressure for industrial engines of this type, the power objectives of 5.2 kW per cylinder for the air cooled and 6.7 kW per cylinder for the water cooled, could be achieved. These ratings being at 3000 rpm for continuous rating.

■ Crankcase

The major component in any engine in terms of size and complexity of casting, machining and investment is the crankcase. In water-cooled engines this normally includes the cylinder water jacket and the crank saddle, whereas air-cooled engines usually have separate cylinder barrels mounted on the crankcase. In both cases separate sumps and mounting faces and a considerable number of machined faces, joints and long oil galleries are necessary.

In the new engines, following much development in the company's own foundry, a totally new concept, involving integral cylinders and sumps, was employed on both air- and water-cooled crankcases. This concept allowed the design to exhibit major advantages in cost and quality by:

(1)   A reduction in the number of machining operations required on the crankcase, which minimized the investment and the component cost.

The crankcase machined faces were reduced from eight to five on both air- and water-cooled versions.

(2)   The common machining of both air- and water-cooled blocks allowed the machining process to control the head face position to within ± 0.12 mm. This allows accurate control on dead volume.

(3)   The reduction of potential leak paths of the crankcase created the opportunity to pay even greater attention to the engine sealing. This feature also helps the quality of the product during its assembly processes.

(4)   The integral sump has shown an advantage in the noise transmitted from the crankcase. The stiffness of the structure, together with the strategically cast-stiffening ribs, has also had an impact on noise reduction.

(5)   The water jacket of the crankcase was designed with casting quality in mind. This has shown a major benefit in the cleanliness of the casting from sand.

(6)   The machining of both water- and air-cooled engines is common. This ensures an improvement in quality as a result of reducing the number of machine settings.

(7)   The crankshaft tunnel design has also given significant benefits in noise and vibration reduction.

(8)   The total crankcase design feature has shown an acceptable weight limit compared with that of the standard design.

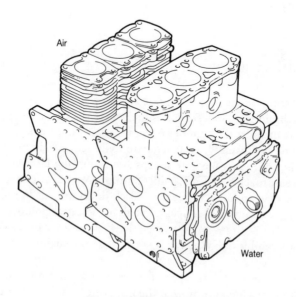

Figure 2 Crankcases.

■ Cylinder head

Another high–cost, high–investment component is the cylinder head. This is invariably machined, not only on the cylinder and rocker cover faces, but also on the side faces which carry the intake and the exhaust manifolds. In the Alpha Series the manifold mounting faces have been eliminated by bringing the port faces out on the cylinder head rocker cover face (Figure 3). This design feature was included on both air- and water-cooled heads, resulting in the following benefits:

(1) Common usage of intake and exhaust manifolds, resulting in reduced cost of complexity and improving equality.

(2) The extended intake and exhaust ports have also shown an advantage in port design, where the upper part of the port functions as the manifold. This reduced the sensitivity of the engine breathing to the manifold designs.

(3) The integrated casting of the ports has shown that various manifold designs can be adopted to the engines, without major effect on the port flows and swirls. This ensures engine performance stability in different applications.

(4) The design of cross–flow head has also enhanced the engine breathing. This allowed optimization of the port shape for both the intake and exhaust.

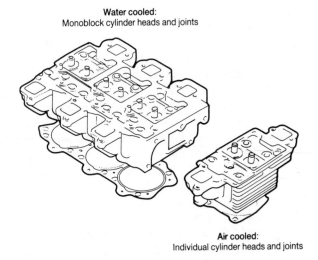

Water cooled:
Monoblock cylinder heads and joints

Air cooled:
Individual cylinder heads and joints

Figure 3   Cylinder heads.

(5)   The water-cooled cylinder head is designed with a one-piece water jacket core. The quality improvement, as a result of eliminating the core assembly processes, is evident on the casting.

(6)   The common mounting of the head for both air and water cooling has ensured the use of four bolts per cylinder. This improved the sealing of the head gasket for both air- and water-cooled engines.

(7)   The valve seat was processed by using the valve guide and the finishing process of both seat and guide was carried out simultaneously. This ensured the concentricity of both features. The process has shown major advantages to eliminate valve seat wear and valve distortion.

■ Crankshaft

The crankshaft is another major cost component and on this range of engines finite-element analysis techniques were used to optimize the crankshaft designs so that they could be fully form cast in SG iron. These components are assembled with cast intermediate bearing housings, and inserted into the crankcase tunnel bore, the housings being located by hollow dowels which carry oil to the thin wall copper lead half-shell bearings. This design fulfilled all of the objectives of low cost, ease of assembly and serviceability consistent with a rigid assembly and minimum bearing clearances, all designed to reduce noise from this source.

■ Cooling system

The cooling system adopted for the air-cooled engine is the traditional use of an axial fan. This permits the optimization of the fan-cooling flow and its distribution to the cylinders.

The cooling system of the water-cooled engine uses a volute-type pump to ensure that required water flow is achieved. The water pump is designed to deliver the required flow for two, three and four cylinders, on both direction injection, DI, and indirect injection, IDI, engines. The volute design also ensured the quality of the flow and the reduction of pump cavitation at higher speed.

The cooling system also has the full-flow bypass feature. This feature will ensure the protection of the engine at any operating condition. It also allows the radiator and fan assembly to be optimized for size and capacity.

The coolant distribution throughout the engines was optimized using transfer holes of the head gasket, with the aid of a plastic cylinder head and flow visualization techniques. The coolant flow being distributed equally across the critical region of the head, that is, valve bridge, and valve seats. This design technique has eliminated any risk of thermal cracking and valve seat thermal distortion for both DI and IDI heads.

The IDI cooling to the valve bridge was improved by the addition of a valve bridge cross-drilling. This being introduced in order to ensure adequate coolant speed in the critical regions of the head.

■ Valve and gear train

The noise emitted by the valve/gear train is always evident in engine designs. In the Alpha Series particular attention was paid to this aspect:

(1) The design reduced the number of gears in the train from five to three (Figure 4), which minimized the mechanical noise of the gear train. Additionally high-quality gears could be used without making an impact on the engine cost. This feature of the engine, alone, has made a major contribution in the reduction of the mechanical noise of the total engine.

(2) To improve the quality of the gear train, together with the reduced investment, the total gear train is mounted on the front face of the block. This improved the alignment of the assembly.

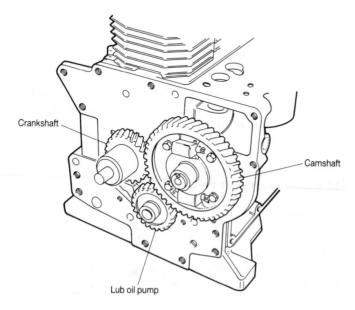

Figure 4 Gear train.

(3) The use of hydraulic tappets in the engine has also eliminated the valve lash, symptomatic of conventional tappets. This reduces the mechanical noise of the train and eliminates the need for valve lash adjustment, which in turn reduces maintenance and assembly costs.

(4) The hydraulic tappet provides a fail-safe device at very high angles of inclination.

■ Lubrication system

The lubrication system is designed to ensure the durability and reliability of the total engine by:

(1) The use of a crank-driven oil pump. The pump capacity is defined to satisfy the requirements of both air- and water-cooled engines with two, three and four cylinders.

(2) The oil pump rotors are sintered to ensure the quality of the assembly.

(3) The lubrication system aeration was reduced to below 1% in order to ensure the operation of the hydraulic tappets.

(4) The relief valve is a non-recirculating type, to reduce the system aeration.

(5) The oil transfer galleries are situated in the side cover of the block. The cast open galleries ensure the cleanliness of the system in relation to that of the long drilled galleries. It also provides equal pressures to the main bearings.

(6) The oil filter is mounted on the side cover ensuring a minimum of pressure drop in the system.

(7) The suction pipe was designed with a carefully optimized suction position to permit a high degree of engine inclination, required on many industrial applications.

■ The combustion process

The design objectives clearly identified the need for low cost, low noise, minimum maintenance and the desirability for common components.

Having considered the very latest techniques of combustion processes it was immediately obvious that to optimize on just one 'state of the art' IDI or DI combustion system would create a conflict in satisfying all of the objectives. It was therefore decided to provide both options for the water-cooled engines. The DI system having the well-known advantages of longer oil drain periods, improved fuel consumption and lower manufacturing costs, and the IDI system having the alternative benefits of lower noise and exhaust emission.

In developing the DI system on both the air- and water–cooled engines, very low combustion noise levels were achieved by sophisticated helical inlet port designs, relatively deep re-entrant combustion bowls and optimized compression ratios, with a fuel injection rate optimized to give short ignition delay periods. However, this very quiet combustion system exhibited white smoke at 'no load' conditions over the speed range up to 3600 rpm and was considered unacceptable for the market.

It was therefore necessary to open the combustion bowl diameters on both engines with a corresponding shallower form, to maintain the optimum compression ratio. This change, unfortunately, raised the engine combustion noise slightly, but eliminated the white smoke by reducing the fuel spray impingement on to the bowl wall. In the interests of lower exhaust emissions this design of combustion chamber was therefore adopted and the small increase in overall engine noise accepted as being the most suitable compromise.

These DI versions of the engine result in good fuel consumption performance of 240 grams/kwh up to engine speeds of 3600 rpm consistent with the lowest manufacturing costs and with oil drain periods of up to 300 hours. However, as engine noise level is such an important criterion, IDI combustion system versions have also been developed for the water–cooled engines.

The design of the IDI combustion system is of the classical type commonly used in small automotive engines. It provides for the quietest combustion level with today's commercially available technology, and is up to 3 dB quieter than the DI system, depending upon the engine speed and load conditions. However, it is more expensive to manufacture due to the added costs of machining the combustion chambers in the cylinder head, and the additional costs of the chamber inserts and the heater plugs required for good starting characteristics.

■ Fuel injection equipment

The fuel injection equipment is one of the most expensive parts of any diesel engine, consequently these components were considered in great detail in order to arrive at the most cost-effective solution.

The first essential design decision was whether to use block type of fuel pumps or individual pumps for each cylinder, or a pump injector for each cylinder.

The block-type pump was an attractive option, as it is one assembly that could be readily attached to the engine with its own governor assembly.

However, to drive the pump would require an additional drive system, either by gear or belt. An extra gear and associated bearing system was established as being expensive and a belt-driven system was not considered durable enough for an industrial diesel engine, consequently the single-pump concept for each cylinder was the best option.

The pump injector was investigated at some length but eventually rejected, as it necessitated a relatively complicated linkage to the engine governor which was not considered viable on these small engines if the precise governing requirements were to be satisfied as demanded on modern generating set applications.

Also the drive to the pump injector concept presented major design problems on these small engines as it necessitated either a push rod from the camshaft or a direct drive from the cam with an overhead camshaft. Driving an overhead camshaft arrangement by gear or belt was rejected for these industrial diesel engines on grounds of cost and reliability. Similarly to provide an extra push rod was both expensive and difficult to accommodate.

It was also important to remember that market requirements dictated a high degree of common components, consequently there were considerable advantages in using single pumps for each cylinder, particularly as the engine range was for two-, three- and four-cylinder versions, air and water cooled, with DI and IDI options. By careful development one common pump has therefore been used for each option whereby considerable advantages have been gained to reduce the number of different parts and consequently reduce overall costs and subsequent spare parts inventory control.

Having decided to use this design concept, investigations then took place to establish the lowest cost, consistent with the highest quality of fuel injection pump and injector. Discussions were initiated with the company's regular European supplier, Lucas Bryce, to establish jointly how best to satisfy the objectives and achieve the cost targets.

Following these discussions, it was decided to adopt a new fuel pump design, using the experiences from Lucas Bryce on their well-proven pump injector range, which eliminated the traditional separate element barrel, thereby saving costs. Also, there was the need to eliminate the necessity of balancing the output of the pumps, within the engine, by adjusting the fuel pump control rack, which on current engine designs creates problems. To achieve this requirement it was agreed to adopt a factory-calibrated fuel pump concept which enabled the fuel pump to be fitted to the engine without the need of any adjustment (Figure 5), thereby greatly simplifying the maintenance and assembly techniques and reducing overall costs consistent with improved quality.

Similarly, the fuel injectors were examined at some length, and by simplifying the design enabled manufacturing processes to be minimized, thereby reducing costs, as well as improving quality. The common injector for both air- and water-cooled versions is also advantageous in these respects.

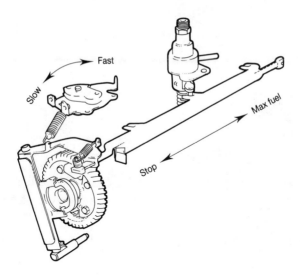

Figure 5   Fuel pump and control rack.

■ Other engine components

There are, of course, many other engine components throughout the engine range which have not been mentioned in this paper, which are essential for operation of the engine. These include such noise-sensitive items as inlet and exhaust silencers, and cooling fans, etc. as well as the numerous engine accessories which are requested by customers to suit various applications. Throughout the engine design programme the company procedures, previously discussed in this paper, were adopted for these components to ensure their effectiveness with respect to noise, reliability, etc. as well as being of minimum cost consistent with manufacturing integrity to ensure the high level of quality expected by the customer.

■ Conclusions

The new Alpha Series of multi-cylinder diesel engines, covering the power range of 8–30 kW, discussed in this paper, designed and developed by Lister–Petter and recently introduced to the market is the only engine range offering both air and water cooling, with DI or IDI options, in the world today (Figure 6). It is believed that the total company involvement adopted by Lister–Petter has satisfied the high demands expected by the customers for

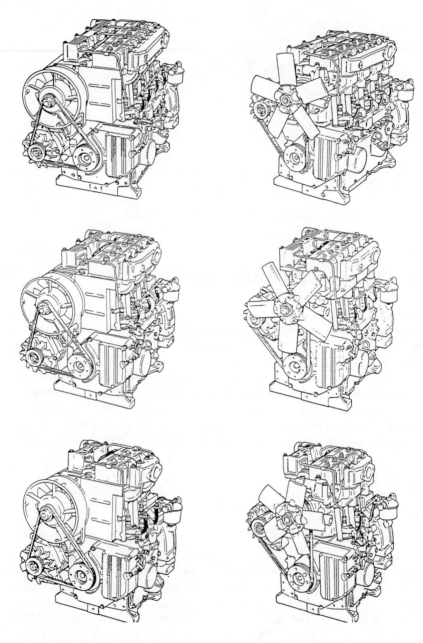

Figure 6   The Alpha Series of multi-cylinder diesel engines.

minimum cost, high quality and performance products.

This paper has outlined the methods adopted within the company to meet the market demands, which have been very successful and resulted in the introduction of the new range of engines in a very short period of time.

The 'design for manufacture' concept adopted on this entirely new range of engines has demonstrated that by simplifying the machining and assembly processes the cost of the final product can be significantly reduced.

These concepts have also demonstrated that by designing for low cost, both the quality and the reliability of the product can be enhanced.

5.2
Creating the 'NPD cell'

James Fairhead
London Business School

The new product development cell is small. It is close-knit and shares a strong sense of purpose. There is a fair degree of overlap between team members, which further enhances its cohesiveness. It is also 'organic' and constantly adapting to different stages of the development process. For all these reasons, the biological metaphor is appropriate. Above all, though, the NPD cell is intensely interdisciplinary and also flexible. In order for an NPD cell to work, a number of basic conditions have to be met.

■ Small numbers and flexibility

An appropriate number for the earliest stages of the new product development and design process may be no more than *one*. All the more so if the product concept or technology is entirely new, since new concepts may initially lack the credibility required to attract support. Gradually, team members can be added, at first, perhaps, *informally*, so that an effective group can be worked out by painless trial and error. Then, when the project receives official approval, it is ready to go flat out towards its objectives, without having to go through the difficult and unproductive 'forming, norming, storming' stages of the 'group development' process.

Where products and technologies are more conventional and objectives widely agreed, it may be possible to cut out some of these transitional stages and go straight to a formalized project team. But there are still strong arguments for letting it form itself. In any event, numbers still need to be kept small. Professor Quinn's US-based study of effective innovation practices suggests that effective interdisciplinary teams need to be composed of

This paper is reprinted by permission of HMSO on behalf of the National Economic Development Office from *Design for Corporate Culture* by James Fairhead. Crown copyright © 1987.

between five and seven members and no more. Part of the reason, he suggests, may be that as a team increases in size from five to eight people, the number of possible one-way communication channels increases dramatically from 75 to 1016 (Quinn, 1985).

But while the core members of an NPD cell are best kept small, it is essential that they include representatives of key functional areas. A common mistake, for example, is to overlook the importance of functions like production engineering and purchasing, because they are 'non-creative'.

☐ Who to include?

But often (as Sinclair knows to its cost with its miniature TV), a key strategic factor is not just the distinctiveness of a new product, but its makeability. It is easy to lose the all-important first-mover advantage if production considerations have not been properly incorporated into its design. Similarly, it can happen that products are designed to international specifications only to find that local component suppliers are unable to meet such high demands. Consultation with the purchasing function at an early stage in the design process can help avoid the poor initial quality and/or dangerous lead-time slippage that results.

This is way the composition of the NPD cell changes continuously over time, even after development go-ahead. The key nucleus will ensure that there is continuity of commitment to the project, and a unification of the key elements that need to interact and be traded off against each other. The nucleus might therefore consist of a product engineer, an industrial designer, a marketer and a production engineer (Figure 1). With these members

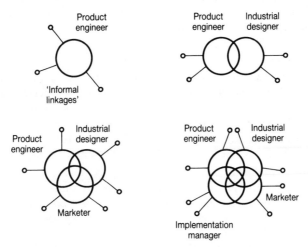

Figure 1   The NDP cell showing one possible development route over time. Each cell is characterized by overlapping responsibilities and 'informal linkages' to functional departments and 'sponsors'.

included, most other functional linkages are covered and there is sufficient overlapping of competence *within* the group to ensure internal checks, balances and integration.

Sometimes the core group will pull in temporary team members to help resolve individual issues or problems. But often, core members will need to form *subgroups* in order to work through particular issues in detail, without obscuring the main thrust of the core group.

When it works well, this is tremendously efficient. The NPD cell reaps the benefits of both co-operation and specialization. But it is also quite complex and it is difficult to write down on paper. It is therefore more difficult to control from on high and requires a greater degree of trust from senior management. But the alternative of designing products in committees of 10 people or more, which many companies continue to adopt, just to play safe, is increasingly untenable. It is often, as we shall see, inefficient, demoralizing and ineffective.

■ Independence

The performance of some cell members (for example, industrial designers, product engineers) depends a lot on their ability to keep up to date with technical skills in their area of expertise (Galbraith, 1971). This could, therefore, be an argument for sticking to traditional 'functional' structures, especially in the case of quite complex technologies. But these are not always good at resolving the numerous conflicts of interest that arise in design and NPD between solutions that are best or most elegant in technical terms, and solutions that are optimal in market terms.

The pioneering BMC transverse engine in the late 1950s suffered from precisely this sort of conflict. Its transmission technology, as Ford knew, was extremely problematic. But the cultural dominance of the BMC engineering department and their blind commitment to this one technology allowed mundane considerations of 'feasibility in use' to be completely ignored. The problem was never, in fact, properly sorted out. Reliability, reputation, sales and profitability were of course the casualties.

In circumstances of conflicting requirements like these (which are all too frequent in times of rapidly changing social, technical and market environments) it is only when multi-functional teams are *at least* semi-autonomous and relatively free from pressures to grind the departmental axe, that a genuinely interdisciplinary process of give-and-take can come about, without causing undue conflict. The advantage of the independent NPD cell approach is that the pros and cons of any particular course of action are more likely to be settled with due regard for broadly *commercial*, rather than narrowly *functional*, considerations.

It is largely for this reason that companies like Ford and JCB are moving over to this way of working, which has long been the practice at high-technology companies like Hewlett-Packard and 3M. However, there is still considerable resistance to this idea among many large UK companies, because it breaks uncomfortably with the established order of things (Edge, 1985).

■ Preserve links with the mainstream business

But despite this need for independence, the NPD cell evidently can't exist in isolation from the rest of the company. In the early stages of its life, especially, it needs a lot of support and encouragement. Later on, when it has got the official development go-ahead, it will still often need to call on central resources for technical assistance, however completely it is staffed up. In short, then, ways have to be found of preserving its network of contacts and influence within the organization, despite its lack of formal command of outside resources. Partly, as we shall discuss below, these links will be preserved by the individual efforts of product champions and product sponsors. But it is also important that top-level leadership does everything possible to create an atmosphere that is *conducive* to this sort of co-operation. In particular, the way that old functional structures are dissolved to create new ones has to be very carefully managed. For there is a danger, in tribal organizations, that NPD cells, once instituted, may be seen as just another tribe, competing for resources and influence, and even diminishing the importance and prestige of the traditional functional departments. This will evidently not encourage the sort of co-operation that new and difficult projects often require.

Creating independent teams can demoralize even the most robust and proud of functional departments. When they lose some of their best young managers to product teams, often in the well-meant cause of 'market responsiveness', they may well infer an unspoken corollary: 'you failed us'. At best, they adopt a hang-dog expression, lamenting the bone that has so unjustly been taken away. At worst, they become distrustful of the influence of the new entities, and cover up their lack of co-operation with a tissue of unproductive myths about the 'incompetence' of these teams.

Some companies, like Ford of Europe, seem to have accomplished this transition relatively smoothly, and productively.

□ The changing role of the functional departments at Ford

At Ford, a gradual and largely informal development of product teams has been going on for some years. For the design department, this at first involved a marked increase in influence. Uwe Bahnsen encouraged his designers to play an increasingly important role as integrators of many previously disjointed aspects of the design process, and they were directly and uniquely under his control.

Gradually as time went by, the cross-functional teams became more and more close-knit and began to develop an identity of their own, partly independent of their functional bases. The development of this autonomy was hastened by the strategic need to shorten lead times significantly from the 52 months it took to develop the Sierra. As part of this, resourcing and operating decisions have been increasingly devolved to the product teams. Members of these teams now work full-time in specific 'programmes' and are responsible on a day-to-day basis to programme directors, not to functional bosses.

But the role of functional bosses and their departments is not restricted to the provision of services, even though they are recognized as an important resource in this respect. The design department, for example, has an over-arching role, which is strategically even more important. This is its role of 'maintaining design standards'. This is taken very seriously, and considerable authority and resources are invested in the design director to achieve this. This involves renewing and building up the professional capabilities within the department and monitoring the quality of their output. The intended technical result is two-fold. Firstly, that design managers have an impressive array of talent to call on for their project teams. Secondly, that they can't get away with compromising on quality and image because of short-term operational pressures. For Ford, this is a terribly important strategic issue.

But there is an important cultural consequence, too. Everyone knows that this new role is important. A far-reaching change in structure has therefore been achieved with minimal disruption and maximum commitment by all parties.

Ford appears to have managed this structural transition rather well; at least in part, this is because it has been taking place over quite a long period. It is also a relatively successful company.

But how should *unsuccessful* companies achieve this sort of change in their design and NPD approach? These companies have often prevaricated for so long that they now face enormous pressure to change a whole lot of things radically and fast – not just in design and NPD but in their mainstream manufacturing and distribution systems, too. So what can these companies do to alter structure and culture without simultaneously demoralizing their existing day-to-day operations?

For most companies in this position, beliefs need to be addressed at two levels. At the leadership level, perhaps a good way to proceed is to institute some new and important ritual to celebrate the expertise and creativity of functional departments, while creating a new identity for them that is attractive and productive. Suitable identities might be 'frontiersman', 'technological guru' or (as at Ford) 'inspirer of technical standards'. (Again, though, to command credibility, this sort of ritual needs to be backed up continually by the words and deeds of important leadership figures.) But beliefs need to be addressed at the individual level too. Firstly, to avoid undesirable élitism on the part of product team members, it may be desirable to make these appointments relatively temporary. But this will all depend . . . some functional specialists with an entrepeneurial bent may best be *permanently* attached to different NPD cells. This need not cause problems, especially where technology and technical skills are relatively easy to keep up to date with.

But another possible device is the 'two-bench' approach whereby team members have two bases. PA Technology follow this practice. It was also described to me in these terms by a marketer in a consumer goods firm: 'After leaving the UK marketing department to take up a staff appointment in international strategy, I often needed access to market and research data about UK operations. Initially there was some banter about my being a "spy for International" but there was no real problem about getting the information I needed. Most of the time I just used to go and help myself. But as soon as my old desk got taken up, it got much more difficult. Partly because I had nowhere to go to do my research without causing some disruption. But it was more than that. I began to feel uneasy about intruding, and began to feel more and more like a nosy staff man. It certainly made me feel very much less inclined to ask favours from market researchers – it was as if I no longer belonged. Maybe they felt that way too. I'm sure that, in a way, the quality of my work suffered.'

The consequences in this case may not have been too significant, but in NPD projects, where the assimilation of quite complex information is at a premium, it may help a lot to adopt the two-bench approach. And this is talking in purely technical terms. If something is made of this device, it can be invested with a lot of useful symbolism. This way, it will have important cultural consequences too. For example, having a well-equipped 'visitors' bench says a lot for the sort of openness and co-operation which is expected of people – certainly much more than a bit of paper urging company unity.

■ Find a 'product sponsor'

A receptive general atmosphere will obviously help a lot, but the NPD cell will need much more. Especially in the early days, and especially if its brief is radical or contentious, it will need advice, encouragement, political support and the occasional helping hand. In many innovative companies, this is the role of the 'product sponsor'.

This is a demanding role. Sponsors, to be effective, need to be experienced enough in the ways of corporate life to know their way around the jungle and to have fought a few battles of their own in their time. They need to be sufficiently part of the establishment to command respect and influence, without being afraid to stick their neck out if necessary (Roberts, 1977; Pinchot, 1985; Quinn, 1985). The job is sometimes so wide and demanding that it is effectively divided into two. Quinn calls this the 'mother' role and the 'father' role, roughly corresponding to moral support and political/financial support.

But regrettably, as Roberts notes, the sponsor's role is often so subtle that its importance goes unrecognized. At companies like 3M, by contrast, sponsorship has been made an *explicit* part of innovation culture – it is a recognized and respected role for ex-product champions. It is therefore less surprising that people can be found to take on this role, even at an age when traditional corporate cultures encourage more relaxing pursuits (Hunt, 1979).

But this is not just applicable to large companies. Perhaps especially in smaller private companies, where relationships between managers and owner–directors can be particularly intense (Scase and Goffee, 1984), there may be a pressing need for this sort of mediating influence.

■ Find a product champion

But a successful NPD cell is not just a popular cocktail of technical experts with a dash of creativity. (Indeed, there is evidence that in the US, at least, too much encouragement is often given to the creative aspects of the earliest NPD stages (Roberts, 1977).)

Instead, there is a need to emphasize the *entrepreneurial* role in design and innovation. In a high-technology company, the role of the corporate entrepreneur may often be to take a bare technological concept and turn it into a product concept with potential use and profit. A classic example of this is provided by 3M's Post-it notes.

□ The story of 3M Post-it notes

The inventor of Post-it notes, Art Fry, reportedly got the idea while singing in a church choir. The slips of paper that he used as hymn-markers kept falling out on to the floor. He conceived the idea of a stickable marker that wouldn't damage the hymnal when removed. By chance, some time earlier, the 3M research lab had sent out samples of an adhesive that had the right sort of properties, and this was his starting point.

But Fry was able to get Post-it notes on to the market only by cutting squarely across a number of functional boundaries. Formulation was one problem – production people said that the stickiness of the adhesive could never be controlled. Production engineering was another – he had to build his own prototype machine because production engineers said it wasn't possible. Arriving at a convincing estimate of market demand was another problem – this he solved by making the prototype product available within 3M and comparing usage rates against standard sticky-tape.

By all these means he was eventually able to build up sufficient data to convince the mainstream functions in the product divisions to support and test-market the idea. Mind you, even this was a problem. The original four-city test-market was a failure. It was only when thousands of product samples were given out free, in a subsequent one-city test, that consumers began to understand the point of the things (Pinchot, 1985, p. 137).

This is far from being an isolated story, of course. Quinn (1985), Roberts (1977), Maidique and Hayes (1984) and Pinchot (1985) have documented numerous other examples of product champions at work in US high-technology companies. They often have to go through a series of more or less clandestine 'bootleg' stages, picking up support and funding as they go along, before they can finally get the official development go-ahead.

Pinchot also makes the important point that there is often an 'intrapreneurial gap' in company organization between the business plan stage of a new product and the time it reaches its growth stage (Pinchot, 1985, pp. 34–5). It *may* be possible to rely on relatively traditional scientists, researchers, inventors and even planners to *initiate* products. But a potentially good concept may get nowhere unless an entrepreneurial product champion then makes the project his, picks it up by the scruff of the neck and battles it through to the stage where it can be handed over to more orthodox product management.

But it would be a mistake to think that the product champion/ bootlegging concept applies only to high-technology products. It was precisely in this way that Black & Decker's highly successful hot-air stripper came about.

As a result largely of the marketing department's scepticism of the idea, it was not possible to get formal agreement for further development. Nonetheless, a small group of engineers unofficially pooled their spare resources and put together some very rough prototypes, comprised of bits and pieces cannibalized from other machines, crammed into an electric drill-casing. They then put this into a small-scale user test and achieved such good results that the project was officially blessed, made top priority and launched in record time soon after.

But this sort of stubborn opportunism can't be legislated – product champions need to volunteer, rather than being appointed, and they are only likely to do this if the prevailing culture of the organization encourages people to go out on a limb in the way they often have to. Hence the crucial importance of the earlier subsections on the power of symbolic leadership and the use of belief creation and reinforcement rituals – without these, managers are unlikely to be convinced that the personal risk of championing a product is worth it. By contrast, many companies have tried to copy the letter of what well-known innovative companies like 3M do but have completely missed the spirit. They have therefore failed.

But the question still remains, 'where do you actually find product champions in the organization?'. The answer is they can come from virtually *any* functional discipline, providing they have certain intrinsic qualities – broad commercial understanding, energy, tenacity and good interpersonal skills are the key requirements. It is from these qualities that they get their credibility within the organization, without which they are likely to find it difficult to attract the necessary support and co-operation.

Potential champions are therefore rare. The qualities they need are very different from those that usually distinguish the highly administrative 'project manager' or product planner. They are not necessarily even to be found in conventional marketing departments – product managers sometimes lack credibility across the broad functional spectrum, even if they are very good at monitoring brand performance and putting together promotional and advertising campaigns.

This is why it is such a crime to allow only 'commercial' functions to participate in this activity, as some companies do.

Doubtless there *are* many marketers with the right sort of qualifications, but it is significant that many of the shining exemplars in the US literature have been engineers, and industrial designers have also been credited with some impressive successes (Lorenz, 1986).

■ Equality of status and authority

We discussed earlier how company culture can easily become fossilized by the conservative influence of hierarchy and official symbols of status. This is also a great barrier to interdisciplinary working, since it inhibits the trust and motivation which are so important for the achievement of difficult and complex task objectives.

Very often, purely administrative 'project managers' (often armed with wildly impressive PERT charts) get paid more and awarded more in the way of 'perks' than technical specialists, such as designers, engineers and R&D men. This has a terribly insidious effect on a project team's ability and desire to contribute fully, freely and consistently. What's the point in working hard if it's the project planner who always gets to meet the VIPs and make all the important pronouncements? The insult is all the harder to bear if he is relatively weak technically and has nothing particular to recommend him in terms of character or leadership.

This is a pity because technical skills, creativity and entrepreneurship, in the right mixture, are the key strategic factors in many industries. Competent administrators are relatively easy to find. Above a certain acceptable level of competence, one administrator is much the same as another, and any marginal difference in efficiency is in any case far outweighed by cultural factors largely outside his control.

This is well recognized by PA Technology, a multi-disciplinary R&D, design and pilot production operation that works for a variety of clients in a number of international industries.

Apart from being highly successful, it is not a small operation – it is more or less the size of a medium-size firm. This suggests that the way that its specialists are compensated uniquely for their *technical* skills, regardless of the level of managerial responsibility they undertake, is quite transferable to manufacturing and marketing operations. This way of treating technical experts is also a common characteristic of successful high-technology firms in the US (Maidique and Hayes, 1984).

Gilbert McIntosh, the Chief Designer of JCB Research, also points out that in sheer time and manpower terms, it is extremely wasteful to give too much influence in the design process to non-technical decision makers.

If the people at the planning end are too remote from the people at the thinking end, everything has to be explained and every option looked at – everything has got to be cut and dried. It's all too long winded.

This issue of 'managers versus specialists' does tend to take on a particular importance for almost any firm involved in component rationalization across product ranges. This is particularly difficult to carry out and extremely wasteful in time terms when the *ex-officio* arbiter between the interests of different product groups is purely a *commercial* person. He may not be sufficiently technical to make what are often highly 'political' choices between different design solutions. For many firms, as we have mentioned, this is a crucial strategic issue. If it is approached in a traditional way, the danger is that nothing happens.

But there is also a risk involved in this egalitarian approach to NPD. It can easily degenerate into what behaviouralists quaintly call a 'love-in'. Partly, of course, it is the role of the product champion to see that this doesn't happen. But conceivably, he might be part of the problem – his every word may end up being taken at face value by other cell members.

There are therefore two final conditions to be fulfilled if NPD cells are to work. Firstly, all cell members must see the process as *theirs*. Thus they have a right to do almost anything (including causing friction) to get it to work well. They will also be quite prepared to do each other's jobs. Secondly, they need to develop a general understanding and appreciation of the *special* skills of their fellow cell members. Otherwise, this job flexibility can get out of hand. It sounds a difficult balancing act, but managers in even the most traditional of companies seem to take to it with enthusiasm and some success.

References

Edge G. (1985). *SIAD Design Seminar*. London: Design Council

Galbraith J. R. (1971). Matrix organization designs. *Business Horizons*, February

Hunt J. (1979). *Managing People at Work*. London: McGraw-Hill

Maidique M. and Hayes R. (1984). The art of high technology management. *Sloan Management Review*, Winter

Pinchot C. (1985). *Intrapreneuring*. London: Harper & Row

Quinn J. B. (1985). Management innovation: controlled chaos. *Harvard Business Review*, June

Roberts E. B. (1977). Generating effective corporate innovation. *Technology Review*, October/November

Scase R. and Goffee R. (1984). Proprietorial control in family firms. *Journal of Management Studies*

5.3

Link manufacturing process and product life cycles

Robert H. Hayes and Steven C. Wheelwright

Business Administration, Harvard Business School, Cambridge, MA, USA

The regularity of the growth cycles of living organisms has always fascinated thoughtful observers and has invited a variety of attempts to apply the same principles – of a predictable sequence of rapid growth followed by maturation, decline and death – to companies and selected industries. One such concept, known as the 'product life cycle', has been studied in a wide range of organizational settings (Wells, 1972). However, there are sufficient opposing theories to raise the doubts of people like Dhalla and Yuspeh (1976), who argued in these same pages a few years ago that businessmen should forget the product life cycle concept.

Irrespective of whether the product life cycle pattern is a general rule or holds only for specific cases, it does provide a useful and provocative framework for thinking about the growth and development of a new product, a company or an entire industry. One of the major shortcomings of this approach, however, is that it concentrates on the marketing implications of the life cycle pattern. In so doing, it implies that other aspects of the business and industry environment move in concert with the market life cycle. While such a view may help one to think back on the kinds of changes that occur in different industries, an individual company will often find it too simplistic for use in its strategic planning. In fact, the concept may even be misleading in strategic planning.

In this article we suggest that separating the product life cycle concept from a related but distinct phenomenon that we will call the 'process life cycle' facilitates the understanding of the strategic options available to a company, particularly with regard to its manufacturing function.

■ The product–process matrix

The process life cycle has been attracting increasing attention from business managers and researchers over the past several years (Abernathy and Townsend, 1975; Abernathy and Utterback, 1975a and 1975b). Just as a product and market pass through a series of major stages, so does the production process used in the manufacture of that product. The process evolution typically begins with a 'fluid' process – one that is highly flexible, but not very cost efficient – and proceeds towards increasing standardization, mechanization and automation. This evolution culminates in a 'systemic process' that is very efficient but much more capital intensive, interrelated, and hence less flexible than the original fluid process.

Using a product–process matrix, Figure 1 suggests one way in which the interaction of both the product and the process life cycle stages can be represented. The rows of this matrix represent the major stages through which a production process tends to pass in going from the fluid form in the top row to the systemic form in the bottom row. The columns represent the product life cycle phases, going from the great variety associated with start-up on the left-hand side to standardized commodity products on the right-hand side.

☐ Diagonal position

A company (or a business unit within a diversified company) can be characterized as occupying a particular region in the matrix, determined by the stage of the product life cycle and its choice of production process for that product. Some simple examples may clarify this. Typical of a company positioned in the upper left-hand corner is a commercial printer. In such a company, each job is unique and a jumbled flow or job shop process is usually selected as being most effective in meeting those product requirements. In such a job shop, jobs arrive in different forms and require different tasks, and thus the equipment tends to be relatively general purpose. Also, that equipment is seldom used at 100% capacity, the workers typically have a wide range of production skills, and each job takes much longer to go through the plant than the labour hours required by that job.

Further down the diagonal in this matrix, a manufacturer of heavy equipment usually chooses a production structure characterized as a 'discon-

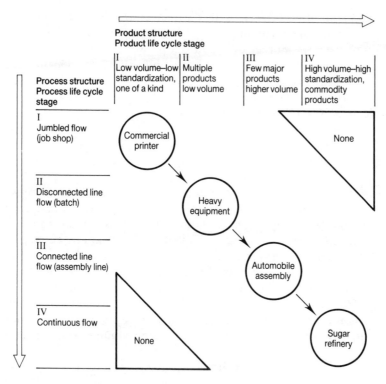

Figure 1 Matching major stages of product and process life cycles.

nected line flow' process. Although the company may make a number of products (a customer may even be able to order a somewhat customized unit) economies of scale in manufacturing usually limit such companies to offer several basic models with a variety of options. This enables manufacturing to move from a job shop to a flow pattern in which batches of a given model proceed irregularly through a series of workstations, or possibly even a low-volume assembly line.

Even further down the diagonal, for a product like automobiles or major home appliances, a company will generally choose to make only a few models and use a relatively mechanized and connected production process, such as a moving assembly line. Such a process matches the product life cycle requirements that the automobile company must satisfy with the economies available from a standardized and automated process.

Finally, down in the far right-hand corner of the matrix, one would find refinery operations, such as oil or sugar processing, where the product is a commodity and the process is continuous. Although such operations are highly specialized, inflexible and capital intensive, their disadvantages are more than offset by the low variable costs arising from a high volume passing through a standardized process.

In Figure 1, two corners in the matrix are void of industries or individual companies. The upper right-hand corner characterizes a commodity product produced by a job-shop process that is simply not economical. Thus there are no companies or industries located in that sector. Similarly, the lower left-hand corner represents a one-of-a-kind product that is made by continuous or very specific processes. Such processes are simply too inflexible for such unique product requirements.

☐   **Off the diagonal**

The examples cited thus far have been the more familiar 'diagonal cases', in which a certain kind of product structure is matched with its 'natural' process structure. But a company may seek a position off the diagonal instead of right on it, to its competitive advantage. Rolls-Royce Ltd still makes a limited product line of motor cars using a process that is more like a job shop than an assembly line. A company that allows itself to drift from the diagonal without understanding the likely implications of such a shift is asking for trouble. This is apparently the case with several companies in the factory housing industry that allowed their manufacturing operations to become too capital intensive and too dependent on stable, high-volume production in the early 1970s.

As one might expect, when a company moves too far away from the diagonal, it becomes increasingly dissimilar from its competitors. This may or may not, depending on its success in achieving focus and exploiting the advantages of its niche, make it more vulnerable to attack. Co-ordinating marketing and manufacturing may become more difficult as the two areas confront increasingly different opportunities and pressures. Not infrequently, companies find that either inadvertently or by conscious choice they are at positions on the matrix very dissimilar from those of their competitors and must consider drastic remedial action. Most small companies that enter a mature industry start off this way, of course, which provides one explanation of both the strengths and the weaknesses of their situation.

One example of a company's matching its movements on these two dimensions with changes in its industry is that of Zenith Radio Corporation in the mid-1960s. Zenith had generally followed a strategy of maintaining a high degree of flexibility in its manufacturing facilities for colour television receivers. We would characterize this process structure at that time as being stage 2. When planning additional capacity for colour TV manufacturing in 1966 (during the height of the rapid growth in the market), however, Zenith chose to expand production capacity in a way that represented a clear move down the process dimension, towards the matrix diagonal, by consolidating colour TV assembly in two large plants. One of these was in a relatively low-cost labour area in the USA. While Zenith continued to have facilities that were more flexible than those of other companies in the industry, this decision reflected corporate management's assessment of the need to stay within range of the industry on the process dimension so that its excellent marketing strategy would not be constrained by inefficient manufacturing.

It is interesting that seven years later Zenith made a similar decision to keep all of its production of colour television chassis in the USA, rather than lose the flexibility and incur the costs of moving production to the Far East. This decision, in conjunction with others made in the past five years, is now being called into question. Using our terminology, Zenith again finds itself too far above the diagonal, in comparison with its large, primarily Japanese, competitors, most of whom have mechanized their production processes, positioned them in low-wage countries and embarked on other cost-reduction programs.

Incorporating this additional dimension into strategic planning encourages more creative thinking about organizational competence and competitive advantage. It also can lead to more informed predictions about the changes that are likely to occur in a particular industry and to consideration of the strategies that might be followed in responding to such changes. Finally, it provides a natural way to involve manufacturing managers in the planning process so that they can relate their opportunities and decisions more effectively with marketing strategy and corporate goals. The experience of the late 1960s and early 1970s suggests that major competitive advantages can accrue to companies that are able to integrate their manufacturing and marketing organization with a common strategy (Skinner, 1969).

■ Using the concept

We will explore three issues that follow from the product–process life cycle:

(1)   the concept of distinctive competence;
(2)   the management implications of selecting a particular product–process combination, considering the competition;
(3)   the organizing of different operating units so that they can specialize on separate portions of the total manufacturing task while still maintaining overall co-ordination.

☐ Distinctive competence

Most companies like to think of themselves as being particularly good relative to their competitors in certain areas, and they try to avoid competition in others. Their objective is to guard this distinctive competence against outside attacks or internal aimlessness and to exploit it where possible. From time to time, unfortunately, management becomes pre-occupied with marketing concerns and loses sight of the value of manufacturing abilities. When this happens, it thinks about strategy in terms only of the product and market dimension within a product life cycle context. In effect, management concentrates resources and planning efforts on a relatively narrow column of the matrix shown in Figure 1.

The advantage of the two-dimensional point of view is that it permits a company to be more precise about what its distinctive competence really is and to concentrate its attentions on a restricted set of process decisions and alternatives, as well as a restricted set of marketing alternatives. Real focus is maintained only when the emphasis is on a single 'patch' in the matrix – a process focus as well as a product or market focus. As suggested by Wickham Skinner (1974), narrowing the focus of the business unit's activities and the supporting manufacturing plant's activities may greatly increase the chance of success for the organization.

Thinking about both process and product dimensions can affect the way a company defines its 'product'. For example, we recently explored the case of a specialized manufacturer of printed circuit boards. Management's initial assessment of its position on the matrix was that it was producing a low-volume, one-of-a-kind product using a highly connected assembly line process. (This would place it in the lower left corner of the matrix.) On further reflection, however, management decided that while the company specialized in small production batches, the 'product' it really was offering was a design capability for special-purpose circuit boards. In a sense, then, it was mass-producing designs rather than boards. Hence, the company was not far off the diagonal after all. This knowledge of the company's distinctive competence was helpful to management as it considered different projects and decisions, only some of which were supportive of the company's actual position on the matrix.

☐   **Effects of position**

As a company undertakes different combinations of product and process, management problems change. It is the interaction between these two that determines which tasks will be critical for a given company or industry. Along the process structure dimension, for example, the key competitive advantage of a jumbled flow operation is its flexibility to both product and volume changes. As one moves towards more standardized processes, the competitive emphasis generally shifts from flexibility and quality (measured in terms of product specialization) to reliability, predictability and cost. A similar sequence of competitive emphases occurs as a company moves along the product structure dimension. These movements in priorities are illustrated in Figure 2.

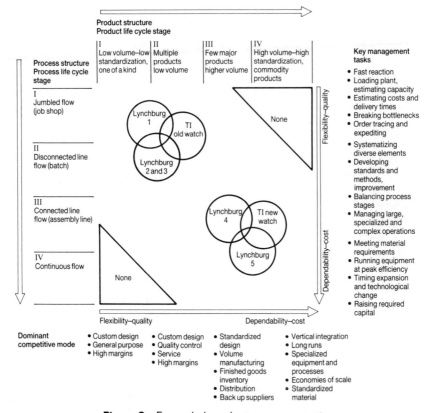

Figure 2   Expanded product–process matrix.

For a given product structure, a company whose competitive emphasis is on quality or new product development would choose a much more flexible production operation than would a competitor who has the same product structure but who follows a cost-minimizing strategy. Alternatively, a company that chooses a given process structure reinforces the characteristics of that structure by adopting the corresponding product structure. The former approach positions the company above the diagonal, while the latter positions it somewhere along it.

A company's location on the matrix should take into account its traditional orientation. Many companies tend to be relatively aggressive along the dimension – product or process – where they feel most competent and take the other dimension as 'given' by the industry and environment. For example, a marketing-oriented company seeking to be responsive to the needs of a given market is more likely to emphasize flexibility and quality than the manufacturing-oriented company that seeks to mould the market to its cost or process leadership.

An example of these two competitive approaches in the electric motor industry is provided by the contrast between Reliance Electric and Emerson Electric. Reliance, on the one hand, has apparently chosen production processes that place it above the diagonal for a given product and market, and the company emphasizes product customizing and performance. Emerson, on the other hand, tends to position itself below the diagonal and emphasizes cost reduction. As a result of this difference in emphasis, the majority of Reliance's products are in the upper left quadrant, while Emerson's products tend to be in the lower right quadrant. Even when the two companies' product lines overlap, Reliance is likely to use a more fluid process for that product while Emerson is more likely to use a standardized process.

Each company has sought to develop a set of competitive skills in manufacturing and marketing that will make it more effective within its selected quadrants.

Concentrating on the upper left versus the lower right quadrant has many additional implications for a company. The management that chooses to compete primarily in the upper left has to decide when to drop or abandon a product or market, while for the management choosing to compete in the lower right a major decision is when to enter the market. In the latter case, the company can watch the market develop and does not have as much need for flexibility as do companies that position themselves in the upper left, since product and market changes typically occur less frequently during the later phases of the product life cycle.

Such thinking about both product and process expertise is particularly useful in selecting the match of these two dimensions for a new product. Those familiar with the digital watch industry may recall that in the early 1970s Texas Instruments introduced a jewellery line digital watch. This product represented a matrix combination in the upper left-hand quadrant, as shown in Figure 2. Unfortunately, this line of watches was disappointing to Texas Instruments, in terms of both volume and profitability. Early in 1976, therefore, TI introduced a digital watch selling for $19.95. With only one electronic module and a connected line flow production process, this watch represented a combination of product and process further down the diagonal and much more in keeping with TI's traditional strengths and emphases.

☐ Organizing operations

If management considers the process structure dimension of organizational competence and strategy, it can usually focus its operating units much more effectively on their individual tasks. For example, many companies face the problem of how to organize production of spare parts for their primary products. While increasing volume of the primary products may have caused the company to move down the diagonal, the follow-on demand for spare parts may require a combination of product and process structures more towards the upper left–hand corner of the matrix. There are many more items to be manufactured, each in smaller volume, and the appropriate process tends to be more flexible than may be the case for the primary product.

To accommodate the specific requirements of spare parts production, a company might develop a separate facility for them or simply separate their production within the same facility. Probably the least appropriate approach is to leave such production undifferentiated from the production of the basic product, since this would require the plant to span too broad a range of both product and process, making it less efficient and less effective for both categories of product.

The choice of product and process structures will determine the kind of manufacturing problems that will be important for management. Some of the key tasks related to a particular process structure are indicated on the right side of Figure 2. Recognizing the impact that the company's position on the matrix has on these important tasks will often suggest changes in various aspects of the policies and procedures the company uses in managing its manufacturing function, particularly in its manufacturing control system. Also, measures used to monitor and evaluate the company's manufacturing performance must reflect the matrix position selected if such measures are to be both useful and consistent with the corporate goals and strategy.

Such a task–oriented analysis might help a company avoid the loss of control over manufacturing that often results when a standard set of control mechanisms is applied to all products and processes. It also suggests the need for different types of management skills (and managers), depending on the company's major manufacturing tasks and dominant competitive modes.

While a fairly narrow focus may be required for success in any single–product market, companies that are large enough can (and do) effectively produce multiple products in multiple markets. These are often in different stages of the product life cycle. However, for such an operation to be successful, a company must separate and organize its manufacturing facilities to best meet the needs of each product and then develop sales volumes that are large enough to make those manufacturing units competitive.

An example of separating a company's total manufacturing capability into specialized units is provided by the Lynchburg Foundry, a wholly owned subsidiary of the Mead Corporation. This foundry has five plants in Virginia. As Figure 2 shows, these plants represent different positions on the matrix. One plant is a job shop, making mostly one–of–a–kind products. Two plants use a decoupled batch process and make several major products. A fourth plant is a paced assembly line operation that makes only a few products, mainly for the automotive market. The fifth plant is a highly automated pipe plant, making what is largely a commodity item.

While the basic technology is somewhat different in each plant, there are many similarities. However, the production layout, the manufacturing processes and the control systems are very different. This company chose to design its plants so that each would meet the needs of a specific segment of the market in the most competitive manner. Its success would suggest that this has been an effective way to match manufacturing capabilities with market demand.

Companies that specialize their operating units according to the needs of specific, narrowly defined patches on the matrix will often encounter problems in integrating those units into a co-ordinated whole. A recent article suggested that a company can be most successful by organizing its manufacturing function around either a product–market focus or a process focus (Hayes and Schmenner, 1978). That is, individual units will either manage themselves relatively autonomously, responding directly to the needs of the markets they serve, or they will be divided according to process stages (for example, fabrication, subassembly and final assembly), all co-ordinated by a central staff.

Companies in the major materials industries – steel companies and oil companies, for example – provide classic examples of process-organized manufacturing organizations. Most companies that broaden the span of their process through vertical integration tend to adopt such an organization, at least initially. Then again, companies that adopt a product- or market-oriented organization in manufacturing tend to have a strong market orientation and are unwilling to accept the organizational rigidity and lengthened response time that usually accompany centralized co-ordination.

Most companies in the packaging industry provide examples of such product- and market-focused manufacturing organizations. Regional plants that serve geographical market areas are set up to reduce transportation costs and provide better response to market requirements.

A number of companies that historically have organized themselves around products or markets have found that, as their products matured and as they have moved to become more vertically integrated, a conflict has arisen between their original product-organized manufacturing facilities and the needs of their process-oriented internal supply units.

As the competitive emphasis has shifted towards cost, companies moving along the diagonal have tended to evolve from a product-oriented manufacturing organization to a process-oriented one. However, at some point, such companies often discover that their operations have become so complex with increased volume and increased stages of in-house production that they defy centralized co-ordination and management must revert to a more product-oriented organization within a divisionalized structure.

■ Strategy implications

We can now pull together a number of threads and summarize their implications for corporate strategy. Companies must make a series of interrelated marketing and manufacturing decisions. These choices must be continually reviewed and sometimes changed as the company's products and competitors evolve and mature. A company may choose a product or marketing strategy that gives it a broader or narrower product line than its principal competitors. Such a choice positions it to the left or right of its competitors, along the horizontal dimension of our matrix.

Having made this decision, the company has a further choice to make: should it produce this product line with a manufacturing system – a set of people, plants, equipment, technology, policies and control procedures – that will permit a relatively high degree of flexibility and a relatively low capital intensity? Or should it prefer a system that will permit lower-cost production with a loss of some flexibility to change (in products, production volumes and equipment) and usually a higher degree of capital intensity? This choice will position the company above or below its competitors along the vertical dimension of our matrix.

There are, of course, several dynamic aspects of corporate competitiveness where the concepts of matching the product life cycle with the process life cycle can be applied. In this article, however, we have dealt only with the more static aspects of selecting a position on the matrix. We will discuss in a forthcoming article how a company's position on the product–process matrix might change over time and the traps that it can fall into if the implications of such moves are not carefully evaluated.

References

Abernathy W. J. and Townsend P. L. (1975). Technical productivity and process changes. In *Technological Forecasting and Social Change*, **7**(4), 379

Abernathy W. J. and Utterback J. (1975a). Dynamic model of process and product innovation. *Omega*, **6**, 639

Abernathy W. J. and Utterback J. (1975b). *Innovation and the Evolution of Technology in the Firm*. Harvard Business School Working Party, HBS 75-18R

Dhalla N. K. and Yuspeh S. (1976). Forget the product life cycle concept! *Harvard Business Review*, January/February, 102

Hayes R. H. and Schmenner R. W. (1978). How should you organize manufacturing? *Harvard Business Review*, January/February, 105

Skinner W. (1969). Manufacturing – missing link in corporate strategy. *Harvard Business Review*, May/June, 136

Skinner W. (1974). The focused factory. *Harvard Business Review*, May/June, 113

Wells L. T., ed. (1972). *The Product Life Cycle and International Trade*. Cambridge, MA: Harvard University Press

5.4
Integrating design and manufacturing to deploy advanced manufacturing technology

John E. Ettlie

School of Business Administration, The University of Michigan, Ann Arbor, MI 48109-1234, USA

Stacy A. Reifeis

Facility Planning and Engineering, Eli Lilly Company, Lilly Corporate Center, Indianapolis, IN 46285, USA

Most managers and engineers in manufacturing have become acutely aware of the outcomes of designing products without the clear anticipation of producing them: they have lost market share to foreign competitors. Because this problem is so well known, we might assume that its nature is well understood or that it is easily solved by purchasing a computer-aided design, CAD, system. In our experience, neither of these assumptions is correct. US firms are just beginning to appreciate the difficulty of interfacing R&D with production. With a few notable exceptions (for example, Ginn (1984)), no systematic empirical research has been done on this issue.

Rubenstein and Ginn (1985, p. 184) addressed the issue of computer-aided design, CAD, and computer-aided manufacturing, CAM, specifically in their review of the R&D production interface literature and

This paper is reprinted by permission of the authors and The Institute of Management Sciences from *Interfaces*, 17:6, Nov–Dec 1987, pp. 63–74 concluded that problems at this interface contribute substantially to the under-utilization of computer-aided design and manufacturing systems. Their conclusion is consistent with the hypothesis advanced by Flynn (1985) that computerization of the design and manufacturing functions in organizations brings existing communication and co-ordination problems to light and exacerbates these chronic difficulties. The case study composite presented by Rubenstein and Ginn (1985, p. 176) suggests that top management's aloofness to a new product's needs for significant changes in processing capability can cause problems at the design–manufacturing interface. These problems may have been just below the surface all along, waiting to be revealed by a challenging co-ordination requirement.

■ The challenge of engineering integration

The share of the world market held by domestic manufacturers has declined steadily in the past five years. Lowering costs and enhancing product performance preoccupies most manufacturing managers, regardless of their locations. Some solutions, like buying from foreign sources and closing plants, are generally perceived as short-run, tactical approaches that buy time for more strategic, pro-active solutions. Much attention is now being paid to the widespread, chronic lack of integration in design and manufacturing so typical of domestic firms.

Major approaches advocated by the design community, and to some extent the manufacturing community, are design for assembly, DFA, and its more comprehensive methodology, design for manufacture, DFM. Traditional organizational structures, information systems and functional specializations simply have not allowed us to focus our co-ordinated energies effectively on reducing costs by anticipating manufacturing in the design. The sequential path a design takes from idea to manufacture has not allowed consideration of product and process simultaneously. The approach Andrew Atkinson (1985, p. 3) advocates is to 'computer simulate both the design analysis and the manufacturing process in parallel'. He cites one case study in which the product development cycle time 'will be reduced as much as 40%'.

A second advocate of design for manufacture, Henry Stoll (1986, p. 1356), sees the problem in broader terms: 'The challenge lies in making the DFM philosophy work under the constraint of existing company policy, organizational structure and administrative practice'. Stoll observes that up to 80% of production decisions are the direct result of product design decisions. Rarely is the production process independent of the product choice and its characteristics – and even when it is, process parameters must be modified to allow efficient operation.

There are a few complete case histories that suggest the nature of this process. For example, in 1969 the Kofu factory of Casio Computer Company was the first in Japan to mass produce electronic calculators. Now it has an integrated, flexible manufacturing and assembly line using 70 robots. The plant manufactures 16 types of calculators and can change from one model to another in less than one minute. Key to the design of this assembly line were 'revisions of all calculator designs [to] reduce the number of their parts' (*Office and Equipment Products*, 1985). Casio personal computers are used to monitor the assembly line; they have completed the design–manufacturing integration cycle by using their own products to produce their products.

■ Case studies of design–manufacturing integration

As part of an ongoing, longitudinal study of the deployment of advanced manufacturing technology in domestic plants (Ettlie, 1985), we compiled case studies of efforts to enhance the integration of design and manufacturing. The performance outcomes of adopting administrative innovations to enhance design–manufacturing integration have not yet been determined. Therefore, these results are preliminary and can best be used to generate propositions rather than to test models of effective integration.

Although a majority of manufacturing firms report that they intend to integrate CAD with CAM in the next five years, few are actually adopting administrative innovations – new policies, practices and structures – to bring this about. In collecting data with interviews in 39 domestic plants, we found only nine cases (23%) where specific, new management philosophies or practices were adopted explicitly to co-ordinate design and manufacturing. We present several of those cases and a few published cases to show the issues and reveal the trends in administration in integrating design and manufacturing.

☐ Off-road vehicles

An off-road vehicle division of a large, diversified manufacturing corporation installed a flexible manufacturing system to machine geometric cast parts that are welded in a nearby robotic cell.

The division did not start out with two separate CAD systems – one for product design and one for tooling and fixturing or processing design – as do most others. They initiated a two-phase program in which programmers did both graphics and numerical control, NC, tape preparation. Numerical control is the technology for storing and automating the instructions to control machine tool motion using punched tape on reels or, in the case of computer-numerical control, CNC, storage is done in computer memory. Programmers are required to have a background in both drafting and NC.

Drawings of parts include shop instructions. Product design and tooling design are done at the same time. At the engineering level, manufacturing engineering and design engineering are required to work together as a 'co-ordinated team'. The in-house computer systems people initially managed the project and were responsible for it. They insisted on one common system – nothing solely home grown was allowed. As a result, a belief in the integration of design and manufacturing has developed, and the old sequential design process was rejected. Product and process design now begin at the same time in the ideal scenario.

The CAD/CAM system is being implemented, with the systems and data processing people in charge. This is in contrast to an alternative, apparently growing trend to create a new position of computer-integrated manufacturing, CIM, czar who rules over both manufacturing engineering and information functions.

☐   **General Electric, steam turbine**

The steam turbine division of General Electric was the CASA/SME LEAD award winner in 1984 for excellence in CIM deployment. Like the Rockwell Space case presented later, the division developed most of the integrating software, but here the structural change for accommodation was at a higher level in the group for deployment. The 10-year program is about half complete and has realized a 35–40% increase in throughput for the manufacturing function; it comes closest to true CIM goals of any published case. In this case, one systems manager headed up a new 'component' called 'advanced automated technology systems' that was created to deploy CIM. This group was created specifically 'to act as an integrating force between manufacturing and engineering', and was composed of five teams:

(1)   advanced engineering and manufacturing for group technology, integrated database and automated process planning;

(2)   advanced systems design responsible for emerging technologies and co-ordinating the master plan;

(3)   systems applications responsible for advanced plant engineering and hardware and software installations;

(4)   NC programming;

(5)   data administration (General Electric, 1985).

☐   **Automotive division**

Since installing an FMS system, a division of an automotive company has initiated a major strategic modernization program in anticipation of new products designed to meet global competition. This division is also planning and implementing significant philosophical change. In the past, it assigned a single engineering designer to a product design project and made cost reduction the responsibility of manufacturing. The division and, in particular, advanced manufacturing planning, have realized that it is virtually impossible to reduce the cost of a product significantly using this old approach.

This automotive division found that it was making design changes in products that were to be discontinued and that it totally lacked an equipment purchase policy or logical method of purchasing equipment. Purchasing just guessed what product design needs might be before the designs were finalized.

Major changes are now underway. More than 100 people will be part of a new product design team effort, including former outsiders like purchasing agents. The number of vendors is likely to decrease, so purchasing has to be involved at some point. A core group from product development will be involved at all times, with other functions like finance and accounting involved for short periods. The goal will be to reduce product cost by 50% using design for manufacturing methods.

Designers will no longer be able to start from scratch because they cannot find old part drawings; they will use an integrated manufacturing database at all times. Specifications for the new product and its manufacturing process will be written jointly by design and manufacturing engineering in an attempt to drop the distinction between the two types of engineering for the program.

☐   **Xerox copiers**

Xerox (Prokesch, 1985, p. 1) '. . . has rethought virtually every facet of its business – from its basic approach to developing and manufacturing products to how it schedules lunch hours for employees'. It has been successful in turning back some of the Japanese competition in copiers.

Xerox uses a complex network of 'product development' teams, 'crisis' teams and 'problem-solving' teams, and it has successfully introduced its 9900 copier in only two-and-a-half years. (It normally takes five years to launch this type of new product.) Since 1980, 'the company has spent nearly $100 million to automate manufacturing and materials handling'. At the same time, Xerox has cut manufacturing costs in half for its $8.79 billion copier business.

Xerox deals with suppliers much as the Japanese do, working closely with 300 vendors instead of the former 5000. Xerox now discourages trying to 'reinvent the wheel'. Only 30–40% of the components of the 9900 copier are unique to that machine. Finally, 'quality problems have been cut by two-thirds' in two years, and based on the index compiled in a monthly survey, customers' satisfaction has jumped 30% during the same period.

□   Tool manufacturer

About five years ago, a professional tool division of a large diversified manufacturing company embarked on a gradual transition to cellular manufacturing. Although upper management agreed to the concept reluctantly, the results of the program have begun to affect planning. Work-in-process inventory has been reduced by 80%; scrap has been reduced by 73%; and floor space has been reduced by 75%. The lead time to ship products from these cells has been reduced from 10 weeks to one week.

In June 1984, a second phase of the cell program began to operate unmanned and with a goal of achieving a 0.5% scrap rate. One of the administrative innovations used to facilitate implementation of this program was a new compensation system. The cell operator's pay at that time was based on the uptime of the second cell rather than on the hours worked or pieces produced.

As the unmanned program was being installed on the second cell, top management changed its plans for the project. During the planning period for the cellular manufacturing program, the division changed its strategy. It owned the majority of the market share of its product line domestically, and it decided to launch a worldwide market program. As part of this effort, it was introducing a new product at a European trade show, and the project was behind schedule. The implementation team for the second cell was asked to produce new parts for the new product to meet the trade show deadline. A joint design–manufacturing development group was established when top management decided to rush the introduction of a new product using the new manufacturing cell.

This joint development team effort also revitalized the effort to install CAD/CAM, a program parallel to the cell program. The electronic data processing, EDP, personnel had participated in planning the installation of the second cell; the new product 'accident' brought them into the CAD/CAM project team. Facilitating this integration was the division's long-time commitment to training systems personnel and other staff, in part through local community college courses in programming. A unique feature of this training was team participation in vendor training. Engineers, foremen and shop floor personnel all attended vendor training courses together, setting up 'family' work groups for program deployment.

□   Appliance division

To overcome the problem of lack of co-ordination between design and manufacturing, a division of a major appliance producer installed a project task force with engineers as project managers to manage production and design for automated assembly of a drive component. When the new system was installed, direct labour savings were 11 people per shift, and virtually no scrap and rework was created by the new system. Changeover time was five minutes for two product types.

In spite of the great success of this new team approach, there were and still are problems in making it work. One problem on the team was 'telling product engineering their job', that is, determining what functional area should take responsibility for what aspect of the project when the outcome is jointly owned. Another problem involved the purchasing department. Special packaging was needed for supplied parts when the system was automated. Purchasing wanted to know who would pay for this special packaging.

The team approach necessitated a new job, called 'automated machine operator'. This person helped to install the new line but was not involved in the purchase decision. In addition to product and process engineers, the team included the maintenance foreman, manufacturing management, quality and sales. A challenge inherent in this case, as in many, was that design engineering was located at headquarters while manufacturing engineering was located at the plant.

☐ Rockwell International space transportation and systems group

This business unit manufactures space shuttle and space station products, producing 1500 parts a week in 800 different configurations in their Downey, California, plant (*CIM Strategies*, 1985). By upgrading an older CAD/CAM system to its fourth application program and third hardware base, Rockwell appears to be approaching a very integrated manufacturing system that most other companies dream of. A total of 53 NC machines are now integrated with CAD.

An interesting aspect of the case is the structural adaptation used: 'NC programmers and shop floor personnel all report to the same person, rather than programming reporting to manufacturing engineering or to tooling', with the full support of top management. The published report stresses 'synergy' and 'team'. In addition, no one has been displaced in the division due to the CAD and CAM integration effort.

☐ Allen–Bradley

Allen–Bradley, headquartered in Milwaukee, produces a diverse line of electrical and electronic components. In late 1984, the World Contactor line based at the headquarters began implementing a 'strategic business objective' (*CIM Strategies*, 1986, p. 1), in order to install a fully computerized line for assembly of a new line of motor contactors in late 1984.

Allen–Bradley's business strategy was totally to eliminate direct labour costs by moving towards full automation. The implementation was carried out by 'planning teams' reporting to a project manager.

Initially, the planning teams consisted of over 25 people representing 'all the departments being affected' (p. 2) including finance, marketing, quality, management information systems, MIS, cost and development. The actual task force formed the design to manufacturing link that proved successful. It consisted of eight members from manufacturing, production, plant, equipment engineering, production and inventory control, test and a special industrial automation system group (acting as the system integrator). The program began in 1983.

Allen–Bradley's project manager had the ultimate responsibility of co-ordinating the implementation itself. He had two main objectives:

(1)   to oversee the building of the line;

(2)   to design the contactors to be made on it.

The concurrent design of system and product took about two years (p. 2).

In the final system, the firm designed and built most of their system machinery. Concurrent with the installation and debugging (approximately six months' time) of the system, the product for the line was being designed.

Allen–Bradley's commitment to the project was strong. In fact the 'tight integration allowed the system to be built faster and Allen–Bradley to enter this new market sooner' (p. 6). The ultimate design allowed them to:

(1)   use no direct labour;

(2)   produce 600 contactors an hour;

(3)   make any of 125 varieties of contactors;

(4)   assure high quality;

(5)   produce and ship orders within 24 hours;

(6)   produce and ship on a first-come, first-served basis;

(7)   be able to produce in lot sizes of one;

(8)   move to stockless production.

The $15 million line now produces 143 variations, and Allen–Bradley aims to capture 30% of the worldwide market share (*Business Week*, 1986).

☐ Amana

In October 1983, a 15-man task force comprised of managers from design engineering, packaging engineering, production control, maintenance, production, manufacturing, quality control and purchasing was assembled to implement an automated production line for Amana Refrigeration, Inc. The task took approximately two years and incurred a cost of about $12 million (*Appliance Manufacturer*, 1985).

Amana began marketing a new series of medium-priced Radar

microwave ranges in April 1985. A task force ensured that the design of the product and the production line were implemented side by side. The two-year program was sectioned into six phases:

(1) planning;
(2) defining requirements;
(3) facility preparation;
(4) checking out initial parts coming off the new tooling;
(5) debugging and fine tuning;
(6) if needed, cost improvements and design changes.

Amana's design for manufacturing team has realized the firm's first application of a transfer tooling line and moved from the traditional conveyor line to the new float (non-synchronous) conveyor (purchased from Hirata of Japan). The float line embraces '1492 feet of conveyor and 65 workstations. For each workstation, cycle time (product in, product out) is four and half seconds' (p. 40).

Amana's task force manager gets much of the credit for the success of this project. Quality has improved immensely, and co-operation and co-ordination between design and manufacturing have been realized. Such co-ordination is now considered essential to all future endeavours. Perhaps some of this success started upstream: 'as need for the new design became evident, so did the need for the new line' (p. 40), said the manager of manufacturing. This initiated a whole new approach to modernization.

■ Emerging patterns of design–manufacturing integration

A number of patterns emerge in these case studies of planned efforts to integrate design and manufacturing. The evolving administrative innovations that are being used to accomplish design–manufacturing engineering are summarized in Table 1. These administrative mechanisms are now presented in greater depth.

☐ Design–manufacturing teams

In three proprietary cases and in three published cases (Xerox, Allen–Bradley and Amana), an unprecedented development team was used to integrate design and manufacturing in the deployment of advanced manufacturing technology. Six of the nine cases used such teams. Some common features are

Table 1 Methods of promoting design–manufacturing integration.

Cases	Design–manufacturing team	Compatible CAD system for design and tooling	Common reporting position for computerization	Philosophical shift to design for manufacturing	Engineering generalists
Off-road vehicles	X	X	X		
GE steam turbine			X		
Auto division	X			X	X
Xerox copiers	X			X	
Tools	X				
Appliance division				X	
Rockwell Space		X	X		
Allen–Bradley	X		X		
Amana	X			X	

notable. First, among the most important participants were the representatives of design and manufacturing engineering, who had been relieved of their day-to-day preoccupation with design change or production problems and were charged with strategic and advanced development work.

Other participants vary: some teams did purchasing; others did not. In one case, the functional department representatives had not sustained their participation in projects and meetings, nor had they delegated authority to their subordinates.

Second, standardization of key design features (for example, bolt holes, rounds and fillets) and a common database for design, marketing, manufacturing, quality assurance and product integrity in service are often taken for granted. Where significant capital expenditures are involved, advanced manufacturing engineering often takes charge of the teams. What happens to this *ad hoc* effort and the degree to which it is co-ordinated, say with a corporate director's function, remains unresolved in these case illustrations. Perhaps many have just not matured to the point where permanent structural changes have been implemented nor are such changes necessarily needed forever.

☐ **Common or compatible CAD system for design and tooling**

Most plants we visit follow an all too common pattern for deployment of CAD technology. A centralized, R&D design function takes full responsibility for product design. Manufacturing engineering located in the plant is identified with production and not with engineering. Greater mobility and status are usually associated with design, and design engineers typically outnumber manufacturing engineers by five to one. As a result, where CAD is implemented to support tooling fixturing, for product changes, it is usually decentralized to plants and not integrated with the product manufacturing design base. Rarely, as in the off-road vehicle case and Rockwell Space, is a common or compatible CAD system used to launch true CAD and CAM integration with local area networks and with a migration strategy to broad band communications technology like manufacturing automation protocol, MAP.

☐ **Common reporting position for computerization**

In four cases, off-road vehicles, Rockwell Space, GE steam turbine and Allen–Bradley, a common, consolidating organizational reporting relationship was installed to integrate design, manufacturing and other functions in the business unit. In the off-road vehicle case, computerization activities were controlled and co-ordinated through the information function in the division. At GE steam turbine, a new group was created, called advanced automated technology systems. Database and process technology, but not manufacturing engineering, report to this manager. In the Rockwell Space case, the new position consolidates activities lower in the unit for NC programming and shop floor. In the Allen–Bradley case, the system designer became the supervisor. The task force manager was successful in staying with the design–manufacturing team long enough to promote a graceful transition from old to new technology in manufacturing.

Two observations can be made about Amana's more or less permanent adapation for accomplishing integration. First, two other cases have come to our attention of CIM-type management consolidation in structural change, similar to the GE steam turbine case. Both have advanced process engineering and information functions reporting to them, and, interestingly, all three are located in the same general geographic area. In one case, the motivation for this consolidation included the desire to separate the information function from the financial section of the business unit as well as the desire to relocate it with a technical manager with R&D management experience. It seems probable that this administrative innovation was diffused locally by the association of the managers involved in these projects. Second, this adaptation illustrates a structural solution to the growing conflict between the management information system, MIS, and the computerized integrated manufacturing, CIM, function.

We know of two more large factory-of-the-future projects where design engineers have been assigned as a liaison to the deployment team. In the first case, the design engineer reports to the plant manager, and in the second case, the design engineer reports to the project manager.

☐ Philosophical shift to design for manufacturing

The appliance division, automotive division, and the Xerox and Amana cases illustrate the growing popularity of resolving the problem of integrating design and manufacturing with a commitment to design for manufacturing, DFM, at higher levels in the business unit or the corporation. One of the major challenges to the implementation of flexible factory automation systems is that the products were designed to be manufactured, assembled, inspected and tested in a partially manned plant. Nearly every flexible automation system installed in the US in the 1970s precipitated changes in the design of the existing or new products to reduce automation costs. These changes were made during start-up, implementation and even after release to production.

The emphasis on design for manufacturing may mean the standardization of fasteners, the elimination of awkward design features, the elimination of processing steps and the substitution of new materials. The engineering movement makes these changes possible, but a philosophical shift precedes their initiation. A significant educational program is a must for this to occur.

☐ **Engineering generalist**

The auto division case illustrates the development of an engineering generalist for advanced planning projects. In another firm, an engineer is moving from design to manufacturing engineering, again illustrating this apparent trend.

We are aware of one additional case involving the division of a major automobile manufacturer that has been discussed at management seminars but not made public. In this case, the division launched a very ambitious automation program with the design for manufacturing effort and a joint venture with a producer of electric motors. Fifty percent of the costs for product parts were saved as a result of this effort. The new engineering generalist for this program was called a 'producibility engineer'.

In another case from our study, a program to enhance the integration of design and manufacturing engineering started with physically locating the two functions together. When product engineers participating in a seminar failed to pass a test on DFM, they were enrolled in a course on DFM. The manufacturing engineers asked why they had not been invited to take the course too – they were thoroughly committed to working with product engineering to enhance product quality and reduce costs, and they felt they should have been included.

All of these cases illustrate the evolution of a new breed of engineer needed to integrate design and manufacturing. We do not know what kinds of individuals fill this role, but they are likely to be quite different from the more narrowly trained and narrow task-assigned engineers typical in many firms today.

■ Discussion

We have discussed five administrative mechanisms used by domestic firms to integrate design and manufacturing that illustrate trends in administrative innovations for deploying computer-integrated manufacturing (Table 1):

- the design–manufacturing team;
- common CAD systems for design and tooling;
- common reporting positions for computerization;
- a philosophical shift to DFM;
- the engineering generalists.

These nine case studies indicate that the DFM and engineering generalist adaptations appear to be recommended regardless of the circumstances. They are likely to be adopted in all successful cases of design–manufacturing integration.

As important as the appearance of these mechanisms for dealing with the integration of R&D and manufacturing may be, there are caveats for their unqualified application. First, little evidence exists that these approaches are more effective or efficient in addressing the problem of integrating design and manufacturing than earlier, more widespread approaches like teams. These administrative mechanisms represent preliminary, emergent, empirical trends and are just beginning to be compared to established management theories.

Second, we have observed these trends among innovative cases of factory automation and CIM in heavy manufacturing. Modernization in the production of durable goods may not be the only model for other situations in integrating design and manufacturing. Finally, a number of these adaptations are driven as much by internal, creative administrative thinking, as by external, environmental pressures from the market place that shorten the product life cycle. These competitive pressures force rethinking of functional specialization for design and manufacturing. Perhaps, the trend towards engineering generalists best illustrates co-ordination in response to these internal and external forces.

Any aggressive manufacturing technology policy calls for new, highly expert manufacturing and advanced manufacturing engineers to address modernization issues in new materials, processing and product design for manufacturing (Ettlie, 1985). Yet, if domestic manufacturing requires 40 000 such engineers in the next 10 years (Kleiman, 1984), existing staff engineers working in other areas will have to be retrained to fill this need.

We have just begun to scratch the surface in resolving problems in integrating design and manufacturing. Questions at all levels of analysis should be addressed (Rubenstein and Ginn, 1985). For example, at the individual level, what types of people can best manage and participate in the integration of design and manufacturing? At the group level, what team organization is most effective with the participation of what functions? Finally, what organizational policies and structures best support the integration of design and manufacturing?

Clearly, engineers should become directly involved in the research addressing these questions to ensure the quality of the results and their usefulness. Both design and manufacturing engineers – defined in broad terms – ought to consider working for advanced degrees in management in order better to contribute to this research effort. The administrative experiments being conducted by many firms on functional integration need the active participation of these two types of engineers and their managers.

References

Appliance Manufacturer (1985). New production line is cost effective for new product. July, 40–3

Atkinson Andrew O. (1985). *Design for Manufacturability: Computer-integrated Design and Manufacturing for Product Development.* SME Technical Paper 851587, presented at the 1985 International Off-Highway and Power Plant Congress and Exposition, MECCA, 9–12 September, Milwaukee, WI

Business Week (1986). The fully automated factory rewards an early dreamer. 17 March, 91

Callahan Robert L. (1986). Manufacturing technology: How much is flash? *Proceedings.* Strategische Investitions Planning für neue Technologien in der Production, University of Passau, West Germany, 5–7 March, 950–63

CIM Strategies (1985). Flexible plans for CAD/CAM in an NC shop. **2**(12), 7–11

CIM Strategies (1986). Building a totally automated line to compete in a new market. **3**(2), 1–6

Ettlie J. E. (1985). *Organizational Adaptations for Radical Process Innovations.* Presented at the 45th Annual National Meeting of the Academy of Management, 11–14 August, San Diego, CA

Flynn Michael S. (1985) Personal communication, 18 March

General Electric Company (1985). *Computer Integrated Manufacturing at Steam Turbine-Generator Operation.* GEC, Schenectady, NY

Ginn Martin E. (1984). Key organizational and performance factors relating to the R&D/production interface. *PhD diss.*, Northwestern University

Hales H. Lee (1986). Design for assembly, *CIM Strategies.* **3**(9) 1–5

Kleiman Carol (1984). An SOS for manufacturing engineers. *Chicago Tribune*, November 25, Section 8, p. 1, col. 2

Office and Equipment Products (Japan) (1985). Efficiency is carried to its limit at Casio's automated calculator plant (abstract from data base: ABI/Inform). **14**(83), 72–5

Prokesch Steven E. (1985). Xerox halts Japanese march. *New York Times*, November 6, Section D, p. 1

Rubenstein Albert H. and Ginn Martin E. (1985). Project management at significant interfaces in the R&D/innovation process. In *Project Management: Methods and Studies* (Dean B., ed.), Chapter 11. Amsterdam: North-Holland Publishing

Stoll Henry W. (1986). Design for manufacture: an overview. *Applied Mechanics Reviews*, **39**(9), 1356–64

Waddell William (1985). *Strategic Management and the Factory of the Future*, presented at the session on computer-integrated manufacturing: organizational problems, at the Fifth US–Japan Automotive Industry conference Entrepreneurship in a Mature Industry, 6 March, Ann Arbor, MI

5.5
Using strategy to create culture

James Fairhead
London Business School

Despite what some management thinkers and consultants seem to be suggesting, the whole apparatus and activity of formulating, monitoring and implementing strategy is *part* of company culture. It is not a purely technical rational/analytic process that stands outside it (Green, 1987). It expresses, more or less clearly, many fundamental beliefs about the nature of the organization and its place in the competitive environment.

And when management thinkers argue that strategy choice is effectively determined by culture they are only partly right. It is certainly true that the influence of something like a strategy committee depends not just on the objective 'rightness' of its prescriptions and the efficiency of its administration and control. For a lot of its effectiveness will depend on whether it expresses and reaffirms beliefs about markets, technologies, capabilities and competition that are broadly *acceptable* to prevailing cultures and subcultures.

But, properly approached, strategy has considerable power to *change* prevailing beliefs and behaviour; it has at its disposal all sorts of subtle ways of 'stretching the envelope' of acceptability. By using symbolism, ritual, mythology and legend, it can legitimize even the most drastic changes in a company's culture and direction. The vehicle for its persuasiveness is its whole panoply of committees, review groups, meetings and documents. It is through these that it spawns many of the stories and myths which shape a company's cultures and subcultures, productively or otherwise.

But therein lies a danger. It is quite possible for strategy rituals to become self-obsessed. Elaborate documentation and complex analysis can easily become an end in themselves. There is something so impressive about strategy documents and operating plans that it is very easy just to do 'more of

This paper is reprinted by permission of HMSO on behalf of the National Economic Development Office from *Design for Corporate Culture* by James Fairhead. Crown copyright © 1987.

331

the same' every year without questioning fundamental assumptions. Managers sometimes know this – they often joke about 'dusting down last year's plan' to save the bother of doing another. But they also know that 'radical' plans are not usually a good idea, so they generally stick to the old routines.

These routines sometimes become enshrined and legitimized by the prevailing culture of the firm. And very often, they are resistant to change, even when they are patently misguided. As Green (1987) notes: 'in general, any belief system is backed up by a disbelief system which defends against proposals or events which threaten its stability'. Later, he writes that 'the strength of beliefs is that they can incorporate conflicting evidence, at least until such time as the weight of evidence forces a paradigm shift (Kuhn, 1970) which happens but rarely. This certainly appears to have been the case at ICI during the 1970s (Pettigrew, 1985).

A typical manifestation of this is the phenomenon of 'groupthink'. Groupthink occurs when a group of people fall into an established pattern of thinking which is bounded and protected by a set of strongly and commonly held assumptions (Janis, 1972). These assumptions are simply not questioned. It is almost inconceivable that they could be. They are part of the fundamental culture of an organization, without which its activities cease to have meaning – unless other, more productive, beliefs are created and put in their place.

Within the corporate world, *any* functional group can fall prey to it, but its narrow determinism is particularly disastrous when it intrudes into marketing strategy and new product development. As Simmonds (1985) notes: 'Many managers, however, seem to be searching for rules-of-thumb about how to compete. From conference room, class and seminar, consultant office and business journal emerge an unending succession of competitive marketing homilies:

- "Never start a price war."
- "Withdraw from any market in which you do not hold a 15% share."
- "In the early stage of the product life cycle, set market share as the key objective and ignore profits."
- "Attack head-on only if you have superior forces."

But competition obeys no inviolate rules. The outcome of a competitive action depends very largely on how the competitors act. . . .'

Yet, particularly in functionally compartmentalized organizations, marketing and strategy groupthink appears to be quite common.

For a variety of reasons, groupthink flourishes best at the functional department level. Against any amount of evidence it stands immutable until a sudden crisis or high-level change of regime drastically recasts policy. The problem is all the less tractable because the whole apparatus of ritual is very often used to *legitimize* this sort of groupthink.

But an effective design and innovation culture is, by definition, open and responsive. It is therefore crucial to guard against the emergence of such a miserable blindfolding of the collective consciousness. A number of successful companies have apparently for this reason built into their design and innovation practices a number of special features. The aim seems to be to institutionalize *challenge and change* into their design culture.

■ Using outside experts

Outsiders often possess (or can be ascribed with) some sort of expert power base. If so, they are in a position to *legitimize* change. This is by contrast with many managers within the company who may for a variety of reasons fail to promote change, even when in a position to do so. For this reason, a company like Olivetti use highly respected (and doubtless expensive) *freelance* designers to work in their offices on important design projects. In this way, among other benefits, they avoid what their design director has called the 'company man attitude' (Thackara, 1982). The power of such outsiders is often considerable, but cannot be taken too much for granted.

A number of external designers have mentioned to me that their presence often allows engineers to do things that they were not allowed to do before. Their viewpoint is legitimized through what might be called a 'myth of superior knowledge'. But this presupposes a degree of respect for designers; their viewpoint can equally be denigrated through a myth of inferior knowledge such as 'you don't know the market like we do'. This is why it is important to choose designers with both technical and interpersonal skills.

There is sometimes another cultural rather than technical reason for using outside experts. Peter Gorb (1985) makes the important point that creativity is sometimes destructive, and often uncomfortable. It poses new and unaccustomed ways of looking at things and is the cause of sometimes undesirable organizational friction. It may, therefore, make sense to 'hire in' creativity, and to 'let it go' when not needed.

This is undoubtedly true of organizations that have inflexible or contradictory sorts of belief. Under these circumstances, creativity may indeed be destructive because it takes place without any guidance from underlying principles. Newness and change and novelty may be exalted too far, without considering wider cultural and strategic issues. But intuitively, it does not seem right to divorce creativity from management completely. Unless managers are brought up to have some personal understanding of the creative processes, how will they recognize the need for it and know how to use these outside experts?

The experience of a number of successful companies, by contrast, seems to show that (while using outside experts) they have built up their own *internal* creative capability, and that this is promoted and guided, very broadly, by the way they have shaped their cultures. A number of mechanisms and myths are promoted which allow for a continuous *simmering* of creative change, rather than forcing creativity 'underground' into some sort of subculture whose time comes, if ever, in a great torrent of scalding steam.

■ Legitimizing 'challenge mechanisms'

Many companies have sought to provide outlets for the constructive resolution of differences of opinion by institutionalizing what might be called 'challenge mechanisms'. These can take a variety of forms.

IBM have legitimized the manager's right to disagree with his boss's decision through a sort of appeal committee. Intel have created a legitimizing myth of 'constructive criticism'. Black & Decker have created a legitimizing myth which says 'you can do anything round here if you can get someone to support you'. An important part of this challenge system is 'bootlegging'. Companies such as Hewlett-Packard, Black & Decker and 3M effectively endorse it by fêting successful bootleg results, and by encouraging the telling of 'legends' about it, which represent it in idealized and stylized form. The setting up of special teams and task forces (often on separate territory, as idealized in the 'skunkwork' legends) can also sometimes work as a symbolic legitimization of challenge and change in NPD and design strategy. A well-known example of this was the setting up of the separate unit to produce the IBM PC. This radically broke with past IBM design and sourcing practices. Another way of encouraging challenge is to legitimize the opinions of groups that are likely to have different beliefs or views. Austin Rover have sought to do this by setting up a 'young people's committee' of younger managers to review design decisions made by the mainstream.

■ The idealization of 'consensus'

Change theorists have long noted that a sense of participation and ownership are important pre-conditions if change is to be welcome and effective. 'Consensus building' is a ritualized activity by which many Japanese firms are able to achieve this sense of ownership.

'Consensus' decision making is frequently used in a pejorative sense in Western business circles. It is believed to mean that decisions have to be unanimous. So the result is either that nothing gets done or that nothing worth while gets done. William Ouchi's (1981) Japanese experience suggests that it is really no more than majority decision making, except with better manners. He identifies three components:

(1)   I believe that you have to understand my point of view;

(2)   I believe that I understand your point of view;

(3)   whether or not I prefer this decision, I will support it, because it was arrived at in an open and fair manner.

The idealization of consensus as part of the strategy ritual should therefore do much to promote the airing of unorthodox beliefs and points of view that may be repressed in other ways of making decisions. Certainly, Sony's creativity and flexibility appear to benefit from this sort of open management approach. It has launched a number of highly innovative products which have turned out to be highly profitable, despite considerable initial opposition within the company (Lorenz, 1986). In many companies without this sort of consensus approach, such product concepts would either not surface because of self-censorship, or would have been rejected by more or less 'democratic' decision making. This consensus approach also accounts for the strength of purpose that Sony management show in the face of difficult new projects and changing conditions generally.

'Once we are convinced about setting up a particular goal, we in the top management of our company devote our entire efforts to the development of the new product, no matter what the cost will be. We believe that these costs are investments in the future of our company. When the product is created, we focus on the production of it without thinking whether it is too expensive to sell when considering its total development costs. At first the product's price may be quite high, and therefore the market may be small. But we will sell it in that small market to establish this new sector of business. Then we will gradually create wider demand until finally we can start mass production for a wider market.' (Lorenz, 1984)

■ Encouraging 'differentness'

Both (2) and (3) above presuppose that unorthodox and challenging beliefs and views do actually exist in a company, and in the strategic stages of design and NPD in particular. It is therefore important to consider how 'differentness' in companies can be stimulated and legitimized without too much threat to the *fundamental* aspects of a company's belief system.

Sony have succeeded in encouraging a degree of diversity within their company, despite the way that Japanese culture militates against it. This is partly through their relatively open recruitment policy, as discussed earlier, and also through their openness to 'new ideas' which a number of visiting Western designers have commented on. Rather like at Black & Decker, new ideas are not threatening – newness is in itself good.

Black & Decker seem to have created something of a 'vive la

différence' myth. This legitimizes differences of opinion between departments at the strategic evaluation and pre-development stage. The story of how marketing originally opposed the heat-gun is told by engineers not as a story against marketers, but as an affirmation of this difference and how it can be constructive. The engineer who told me the story also added that very often engineering gets it wrong. Although fundamental beliefs are largely the same, it is recognized that differences of perspective and emphasis do exist between departments. Their practice of encouraging overlapping competences is another device that tends to promote this constructive sort of diversity.

■ Reducing the insulation of hierarchy

As we have mentioned, the strategy ritual is sometimes the exclusive preserve of the upper echelons of a company. This is often because it is effectively being used to symbolize the achievement by senior managers of status, influence and power. Under these circumstances, there is bound to be a resistance to change and creativity, since anything that causes a departure from past strategy also signals a loss of personal status. (This is often why old strategies and product lines are so hard to abandon.)

By contrast, companies that are strategically responsive in design and NPD seem to have devolved a lot of the strategy-making process. But ultimate control is always likely to remain higher, rather than lower, in the hierarchy. It is therefore still important to see that this correspondence of rank and task does not shut out criticism from below. It is very easy for even the most liberal minded of senior managers to resent criticism and find it threatening, especially if he is palpably wrong.

One way of dealing with this is to institutionalize some sort of myth which explicitly addresses the often uncomfortable question of senior management fallibility. The 'we all make mistakes myth', for example, helps to make it somewhat more palatable for people to accept that they are wrong. At Apple, Steve Jobs (1981) created a somewhat more *robust* version of the same myth, backed up by action, that senior management should be allowed the same 'privilege' that everyone else has – to be a 'jerk' occasionally.

'In the early days Jobs could go to a product development meeting and say "god, that's a piece of shit" and it would be "just the same as if anybody else had said it". Later, as the company got bigger he began to find that people would automatically take his opinion as fact, even when he didn't know what he was talking about. So in his words: "we had to instil in people that we operate in a dual mode as management, and that they can't take away the privilege of us being individual contributors or individual jerks either. . . so we try to hire people who can listen to all the inputs and then go ahead and do what they want to do." '

This recognition of senior management fallibility is also an explicitly declared aspect of culture at Intel.

The Intel myth of 'constructive criticism' very much includes the opinions of senior managers. One of their important large-scale meetings is to review current strategy (Quinn (n.d.)). This is partly a symbolic affirmation of unity, but it is also a powerful and symbolic protector of change. The fact that it is planned and officially sanctioned helps to reduce the destructiveness of even the most bitter criticism, and helps to make it acceptable and acted upon. It is a way of institutionalizing self-criticism and open mindedness among top management, and reducing the stigma of 'being wrong'.

But perhaps the most powerful action that can help reduce the insularity and rigidity of strategy formulation in hierarchical organizations is to attack symbols of status. Many managers in creative and innovative firms believe that they interpose an unhealthy barrier between strategy and reality. Peters and Austin (1985) recount how Zaphiropoulos of Xerox office systems approached the vexed question of reserved parking places: '. . . he bought two one-gallon cans of black paint and a paint brush. He returned to the office. And then, in front of his office building, the chief executive stripped off his coat, rolled up his sleeves, opened a paint can, and while the cameras rolled (the two phone calls had summoned the video people) he began to paint out the labels on the executive parking spots. Symbolism? You bet. "Strategy" in action? You'd better believe it!'

This is far from being just an eccentric phobia of exaggerated significance. Maidique and Hayes's (1984) impressive and extensive study of successful and unsuccessful US technology-based firms confirms the importance of such symbolism: 'A source of division, and one which distracts the attention of people from the needs of the firm to their own aggrandisement, are the "perks" that are found in many mature organizations: pretentious job titles, separate dining rooms and restrooms for executives, larger and more luxurious offices (often separated in some way from the rest of the organization), and even separate or reserved places in the company parking lot all tend to establish 'distance' between managers and doers and substitute artificial goals for the real ones of creating successful new products and customers.'

But there is an obvious danger in focusing attention on all this easily identifiable hierarchical symbolism. Mere tokenism is not enough. Many UK companies have abolished separate dining rooms, and instituted multi-functional committees, for example, but not much else changes. It is important to manage culture consistently, bearing in mind the interrelationships of all three levels that we have discussed. This means addressing authority relationships, attitudes and ways of working, too.

By contrast, at Intel and similar non-traditionalist companies researched by people such as Maidique and Hayes (1984), Quinn (n.d.), Peters and Waterman (1983), and Peters and Austin (1985), such things as the open strategy sessions and lunching arrangements are accompanied by a *genuine* open-minded style and workaday visibility by top management, which is repeated down the line. Things are therefore rather different. Quinn reports of Intel: 'Any of the top three was likely to plop down at a table . . . and chat with whoever was there. Said one group of employees: "it's exciting to know you may see and talk to the top guy at any time, you feel a real part of things . . .".'

■ Encouraging functional equality of status

Ideological flexibility in strategy making is also hindered by differentiated status between the different *functions* of a company. Yet a consistent comment made by consultants, observers and practitioners alike is that engineers, designers and production people are relatively badly paid and motivated in UK firms compared to their colleagues in *commercial* departments. Once again, this is made very visible in 'fringe benefits' such as the comparative luxury of an office or the size of an entertainment allowance.

Under these circumstances, it is unlikely that strategy formulation rituals will do anything else but reaffirm the existing belief systems of particular functional groups. Strategy implementation, especially where it involves anything new, will either be non-existent or half-hearted. Maidique and Hayes's (1984) research confirms this very strongly: 'Product design, marketing and manufacturing personnel must collaborate in a common cause rather than compete with one another as happens in many organizations. Any policies that appear to elevate one of these functions above the others – either in prestige or rewards – can poison the atmosphere for collaboration and co-operation.'

■ Democratic leadership

We have already discussed the symbolic potency of leadership as a way of creating culture, suggesting that while there are all sorts of leadership styles, leadership is effective to the extent that it is able to communicate a coherent system of acceptable beliefs.

A fundamental characteristic of effective design cultures seems to be a belief in the need for openness. This is what enables certain companies to be receptive and flexible in the way they approach strategy. The challenge mechanisms that we have described are an important way of institutionalizing this flexibility of purpose, but their effectiveness rests ultimately on the ability of leadership figures to symbolize this openness and approachability in the way they act.

■ Making strategy effective through symbolism

When top management from companies like Philips, Sony and Honda make one of their regular visits to the design studio, there is, of course, an exchange of views and the direction of the company is perhaps influenced by this to some extent. But much more important is the *symbolic* meaning of the visit. It reaffirms to management that product design really *is* the key aspect of current and future company strategy, and that it is something which needs to be discussed openly and informally. Symbolic actions of this sort have a cumulative and synergistic effect. Eventually, it is to be hoped, the importance of creative and innovatory design becomes a highly motivating *belief* among managers. Design activities then take on a liveliness, purpose and urgency which strategy documents alone cannot bring about.

■ Summary

The strategy ritual is one of the most potent tools available for the creation of new beliefs. The example of a number of innovative companies shows how leaders can enhance the power and flexibility of strategy not only through their own symbolic behaviour, but by institutionalizing a whole system of strategic ritual and subritual that encourages appropriate beliefs and behaviour. The hoped-for result is that companies can be consistently *creative* in their approach to design and innovation strategy, while paying due regard to company and market imperatives.

References

Gorb P. (1985). Bridging the gulf. *London Business School Journal*, Winter

Green S. (1987). Strategy and organizational culture. *Organizational Dynamics*

Janis I. (1972). *Victims of Group Think: a Psychological Study of Foreign Policy Decisions and Fiascos*. Houghton Mifflin

Jobs S. (1981) *Design at Apple Computer*. Conference Paper of the Design Management Institute, Boston MA

Kuhn T. (1970). *Structure of Scientific Revolutions*. Chicago IL: University of Chicago Press

Lorenz C. (1984). A vicious race to get ahead. *Financial Times*, 19th September

Lorenz C. (1986). *The Design Dimension*. Oxford: Blackwell

Maidique M. and Hayes R. (1984). The art of high technology management. *Sloan Management Review*, Winter

Ouchi W. G. (1981). *Theory Z*. Reading MA: Addison-Wesley

Peters T. J. and Austin N. (1985). *A Passion for Excellence*. London: Collins

Peters T. J. and Waterman R. H. (1983). *In Search of Excellence: Lessons from America's Best-run Companies*. New York: Harper & Row

Pettigrew A. (1985). *The Awakening Giant: Continuity and Change in Imperial Chemical Industries*. Oxford: Blackwell

Quinn J. (n.d.). Case Studies

Simmonds K. (1985). Peaks and pitfalls of competitive marketing. *London Business School Journal*, Autumn

Thackara J. and Lott J. (1982). Design management: in search of an agenda. *Design*, September

5.6
Organizing for manufacturable design

James W. Dean, Jr and
Gerald I. Susman

Pennsylvania State University, USA

Nowhere in a company is the need for co-ordination more acute than between the people who are responsible for product design and those responsible for manufacturing. As Daniel E. Whitney argued in these pages recently ('Manufacturing by Design', *Harvard Business Review*, July/August, 1988), most companies have operated for years in an environment where design and manufacturing communicate infrequently, if at all. In the worst instances, product designs were just thrown 'over the wall': designers felt that their job was finished when designs were released and disappeared into manufacturing's domain; manufacturing engineers struggled to build products that were dropped in their laps.

Many companies have come to appreciate the disadvantages of this sequential approach to product development. Final designs emerging from engineering may be producible only at very high cost. While design expenditures *per se* may amount to only a small part of a product's total cost, design determines a huge proportion of producing, testing and servicing costs. Forcing manufacturing to wait to begin its work until a design is released prolongs the development time of a product and may force a company to miss a market opportunity. Often a company is forced to play catch-up by implementing numerous engineering changes long after products have been introduced.

Effective manufacturers work from designs that have as few parts as possible, as many standard parts as possible, and that can be assembled by methods within manufacturing's capabilities. Barely manufacturable designs compromise product reliability and may preclude the use of robots or computer numerically controlled machines whose purchase was justified on the assumption that they could be kept busy.

Companies that have tried to design for manufacture merely by exhorting designers to create more producible designs, that is, without changing the basic organization of product development, have run into serious trouble. There are, after all, barriers to the integration of design and manufacturing. Often engineers in the two corporate functions have had different educations and share neither a common language nor compatible goals. Design engineers tend to be more focused on the product's performance or on its aesthetics; manufacturing engineers generally concentrate on plant efficiency.

In fact, designers generally enjoy higher status and pay than manufacturing engineers. Designers are considered something akin to creative artists and may be rewarded for ingenuity that has little to do with whether or not their designs can be turned into products cheaply and easily. Manufacturing people, whatever their background, often bear the stigma of less well-educated people, managers who have worked their way up from the factory floor.

The personnel from the two functions may also be located in different buildings, cities or even countries, and there may be no real opportunity for them to establish rapport. Budgeting practices may exacerbate rivalries, as manufacturing engineers are often not funded to work on projects until the design is released, which is precisely when funding for design ends.

Having observed numerous manufacturing organizations, we've discerned several organizational approaches to designing for manufacture that go a long way towards overcoming these barriers. They range from manufacturing sign-off on designs, to combining product and process engineering into the same department. Some are more sweeping than others, but all are quite practical, having been used by a variety of companies. Fundamental to all of the approaches is basic change in the *structure* of the organization.

■ Manufacturing sign-off

In this approach, manufacturing engineers are given veto power over product designs, which cannot be released without manufacturing's approval, though in some cases, only its final approval. It is unlikely with this approach that an unproducible or barely producible design will reach the factory floor. But the approach's biggest drawback, clearly, is its heavy handedness: it gives a club to manufacturing without providing for creative interchange between the two functions and does not allow manufacturing to begin its work until design's work is completed.

Companies using this approach seldom let designers grope blindly for what will satisfy manufacturing. Designers can use commercially available software to assess a product's producibility. Among the best known systems are Boothroyd and Dewhurst's design for assembly and Hitachi's assemblability evaluation method. These and other systems calculate a producibility score for nearly any product, based on the number of its parts, the number of its standardized parts, the simplicity of couplers, the motions involved in its assembly, and so on. Some programs can even generate barcharts that demonstrate the contributions of subassemblies to manufacturing cost and time.

The manufacturing sign-off is relatively simple to manage and depends little on interpersonal skills of engineers on either side of the wall. Designers' use of expert system software permits smaller companies to benefit from the accumulated knowledge of more advanced companies and allows manufacturing engineers to focus on their principal task, namely, design of the process.

Incidentally, a number of companies have created customized software for their design engineers. Customized packages aim to give designers information about the specific constraints of the production site. Other companies use less sophisticated (and less expensive) paper versions of the same idea, that is, listings of preferred components, standard production routings, and so forth. These documents are typically distributed to the design group, so they can try to make their designs manufacturable before manufacturing approval is solicited.

One appliance manufacturer we know, with several billion dollars in annual sales, has its product designers use a number of self-checking software modules to monitor such qualities as mouldability (for plastics), formability (for metal drawings) and ease of automated assembly. The company wants products to achieve acceptable scores on manufacturability before manufacturing even sees the designs. This allows designers to see these programs as a tool to help them, rather than something that manufacturing uses to frustrate them. Approval must be granted by a producibility–feasibility group, however, at as many as 10 phases in the design process. If approval is withheld, the design cannot pass to the final phases, when designers would ordinarily order equipment to build prototypes.

Senior managers at this appliance company believe that their software programs have educated designers and that they've helped the company make major strides in cost and quality. Again, however, the system's greatest weakness is the lack of day-to-day communication between designers and manufacturing people. As a manager who helped to design the current system put it, designers still 'work in a vacuum'. He wishes there were more interaction 'when the paper is clean'.

The company has considered bringing design and manufacturing into the same building, though the cost is currently prohibitive. Plans for shifting responsibility for building prototypes from design to manufacturing are underway. Executives are also contemplating letting manufacturing contract for tool making by outside vendors before the design is finalized so vendors can be 'partners' in the process. However, such an approach may lead to complex (and possibly costly) negotiations with these vendors whenever a design is changed, and managers in finance are resisting it.

■ The integrator

Integrators work with designers on producibility issues, serving as liaisons to the manufacturing group. Naturally, such a role requires individuals who can keep design and manufacturing perspectives in balance. An integrator who leans too heavily towards manufacturing will lose credibility with designers, and someone who leans too heavily towards design will simply not get the job done.

Given the way engineers are currently educated and promoted, integrator candidates may be hard to find. Manufacturing and design engineers are the products of separate and distinct degree programs. And once working for a company, they tend to be promoted within their respective hierarchies. There is usually little opportunity to broaden their provincial outlook to include the other group's concerns.

The integrator approach has been used by the electronics division of a multi-billion-dollar company we know, a company that sells advanced avionics and communications equipment to the military. Now that the Defense Department insists that its contractors make provision for the producibility, serviceability and maintainability of weapons systems, the division must pursue design for manufacture even more carefully. Moreover, since it recently spent tens of millions of dollars on an advanced electronics assembly plant, the company wants to see its products designed to take advantage of this new facility.

Company executives originally hired as integrator an industrial engineer with a background in design. He began by preparing a manual for the design engineers to use and by securing top management support for manufacturing sign-off on product designs. The manual, planned as a collaboration between design and industrial engineering, was to be divided into five volumes, covering such things as lists of standard components and circuit board mountings to the more obscure cost implications of producibility mistakes in complex, low-rate products.

Company executives also assigned a test engineer to make sure components were manufactured in a way that allowed for systematic inspection, and they issued a directive that the producibility co-ordinators were to approve all designs before release. As one supervisor put it, 'the preachers became auditors'.

This division has already seen substantial pay-offs from these efforts. Boards on which automated component placement techniques can be used have increased from 40% to about 55%; for newly designed boards, almost 90%. Plans include designating a third producibility co-ordinator, as well as hiring a number of scientists to develop new production methods for the advanced designs that it will soon have to produce. The company also plans to involve its subcontractors in future efforts to improve producibility.

Clearly, the integrator approach is reasonably flexible. A single individual (or a small team) can easily keep track of new capabilities in manufacturing. Manufacturing engineers don't have to become more knowledgeable about design or designers become more expert in manufacturing. Rather the approach develops an 'expert' in producibility, who can become the focal point for company-wide efforts.

There are some disadvantages, however. One problem is the downside of the integrator's virtue, what one integrator we spoke with called the 'guru syndrome'. Since the integrator is there to worry about producibility, no one else does. The integrator approach makes the organization very dependent on one (or only a few) individuals. It does not facilitate simultaneous engineering – that is, manufacturing cannot begin its work before design's work has been completed.

■ Cross-functional teams

Another step away from the traditional approach is cross-functional teams. At a minimum these consist of a designer and a manufacturing engineer, who work together throughout the whole process. The team meets regularly or may even be located in the same office.

This is the first approach that facilitates simultaneous engineering: the manufacturing engineer becomes familiar with the design well before it is released and may even have had a hand in creating it. The manufacturing process can be partially if not completely planned before the design is finalized.

Perhaps the best example we know of the cross-functional team is a process control company with annual sales of about $100 million. It became interested in producibility about two years ago, when, like the military contractor we just discussed, it began to utilize auto insertion equipment for circuit board production. The company was also discouraged by the failure of one of its recent products whose poor quality was attributed to a barely producible design.

Initially, individuals from manufacturing engineering and quality assurance were formally assigned to the design team, and from the first day of a new product program, the entire group met once or twice a week. A program manager, who acted as a mediator, determined the meeting schedule. Eventually people from test engineering, purchasing and marketing joined the team.

The new approach created some frictions. Designers felt that quality assurance and manufacturing were pre-empting them and wondered why the company didn't trust them to create good designs independently. They also felt manufacturing's demands were often unrealistic, particularly concerning clearances among the various components. They were upset that the new system undermined their creativity.

These sentiments have all dissipated over time, largely because the company took the trouble to calculate the relative cost of various designs at the circuit board level and proved beyond doubt the importance of using auto insertable components. Costs are so much lower now with automated assembly that the team recently set a goal of using auto insertable components 97% of the time. Incidentally, while conflicts between design and manufacturing have generally been settled in team meetings, final authority over design rests with the engineering manager.

The team approach has also been used successfully by a multi-billion-dollar aerospace company we've studied that produces for both military and commercial markets. The company wanted to decrease its development time for major products from 36 to 24 months, to reduce the number of engineering changes necessary to meet production standards and to fully utilize its advanced manufacturing capabilities.

The producibility effort began in 1983 with a series of seminars for designers. By 1985, the company was ready to experiment with what it called product centres. It formed design teams with representatives from engineering, manufacturing, quality assurance and product support (documentation, etc.), with either design or manufacturing providing leadership.

The product centre concept has continued to evolve. There are six now working on a single big project, organized around major subcomponents. Each product centre has a manager and deputy manager and, whenever possible, people working on the project are located in the same room. Manufacturing begins process planning as soon as a week or two after design begins its work. Traditionally, manufacturing engineers would not begin to work until designs were released; today they *finish* almost simultaneously with design release.

The possibility of simultaneous engineering is one benefit of the team approach. Also, the frequent interaction involved (especially when product and process designers are located in the same place) permits people from the two functions to educate one another, thus enhancing capability for future efforts. Perhaps the most attractive feature of this team approach is the way it substitutes collaboration for auditing. Yet team members report through separate hierarchies, which helps them to hold to their respective missions. There are tensions, but these often stimulate greater creativity.

The team concept is expensive because it means assigning people to a development effort not only during the time when their special expertise is crucial but before and after as well. The approach requires members to gain broad expertise in producibility, since there is no longer a single producibility expert, and it demands excellent interpersonal skills. There are probably some engineers on both sides of the wall who would do more harm than good on such teams. Finally, the team approach runs the risk associated with any method that allows simultaneous engineering: process planning done before the final design is released may have to be scrapped as the design evolves.

■ The product–process design department

Our fourth approach to design for producibility involves the greatest degree of structural change. It entails creating a single department responsible for both product and process. A number of variations of this approach are possible, including:

- A senior manager with responsibility for both product and process design, but separate subunits for each function.

- Product and process engineers combined into a single department with one manager having responsibility for both groups.

- One department composed of product–process engineers, that is, individuals with responsibility for both aspects of the design, a rarely found ideal, since very few people have the skills necessary to straddle both worlds.

One $100 million defence technology company we know learned the virtues of a product–process department the hard way. At first it subjected products to non-binding manufacturability reviews; this worked fairly well until the value engineering group came up with an idea for reducing the cost of a critical component by substituting a die casting for a machined blank. Manufacturing liked the concept but contended that vendor-supplied components would not perform as promised without costly rework. The value engineers dismissed manufacturing's objections, and manufacturing didn't press the issue.

Unfortunately, the protests were justified. Reprocessing increased the cost of the component by almost 100%, as much as $300 000 over the life of the program. Ironically, the company had already intended to move to the product–process department approach and had even begun giving the head of value engineering the added responsibility of managing the manufacturing engineering department. As it happened, unfortunately, this person was not comfortable with the manufacturing group and spent very little time there.

Now the company is operating with an interim structure. In the absence of an organization that supports manufacturable designs, producibility depends heavily on the personal inclinations of the designers. The company's computer-aided design system is networked between design and manufacturing, and designers can discuss the implications of designs with manufacturing while both view the drawings on their screens. Unfortunately, this happens far less often than management would like.

An automobile components company, a subsidiary of one of the big three, has the second type of product–process department; design and process engineers report to the same manager at the first level. Product–process teams share software systems and offices, and members are encouraged to decide on a design among themselves. If they don't, their manager resolves the issue. The company adopted this approach in order to reduce development time and cost, and these goals have been achieved, in addition to a reduction in engineering changes after products are in production.

The one-department approach permits simultaneous engineering and leads to mutual education through day-to-day contact. It also places a high premium on the technical and interpersonal skills of department members. The department head must strike a balance between engineering functions and bring a great deal of expertise to the table.

The one-department approach creates the greatest degree of structural change of the four approaches. It may also create the greatest resistance, as people are torn from the comfortable surroundings of their professional and departmental loyalties. Perhaps the greatest danger of the one-department approach is that, as the design and manufacturing communities work more closely together, they may find it too easy to compromise on designs that are merely acceptable to both sides – rather than insisting on functional excellence from their own disciplinary standpoint.

It is clear from our examples that a number of different approaches can lead to producible designs. How can managers choose among the alternatives? The approaches range in impact on the organization from manufacturing sign-off (relatively low) to the product–process department (quite high). Higher impact approaches allow for collaboration and simultaneous engineering, while lower impact approaches do not. On the other hand, higher impact approaches place much greater demands on people, in terms of absorbing change and developing new skills, both technical and interpersonal.

Companies that enjoy substantial freedom in product and process design, and that have the organizational ability rapidly to absorb change, are well positioned to take advantage of the higher impact approaches. Such companies can profit from the collaboration and simultaneous engineering that such methods afford.

Companies whose products and/or processes are relatively fixed, or whose capacity for absorbing change is limited, would be better advised to begin with manufacturing sign-off or the integrator approach. When products or processes are fixed, there is little to be gained by the intense interaction that the higher impact approaches feature. And organizations slow to change would take so long to implement the higher impact designs that they would get greater benefit by using one of the simpler approaches. Of course, such companies could always move to higher impact designs at a later date.

These approaches are not cast in concrete or meant to seem exclusive of one another. Structures for organizations ought to accommodate the messy dilemmas managers face, so each organization should customize its own approach, using the four identified here as building blocks. Whatever the chosen method, change in organization structure will be necessary in a program to achieve manufacturable design.

Conclusion

Design for manufacture, DFM, in a general sense, relies on a closer working relationship between the product design activity and manufacturing activity with the aim of improving manufacturing performance. The need today to integrate the two functions in some way has never been more urgent. It is driven mainly by competition with industries striving to reduce costs, improve quality and get products in the market place quicker.

As with other 'performance drivers' such as total quality management, continuous improvement programmes, total elimination of waste or world-class manufacturing, currently operating in manufacturing industry, DFM is a complex approach. Unlike say computer-integrated manufacturing which is usually interpreted as a largely technological solution to integrating different functions within a manufacturing organization (mainly through computer-based communications), DFM is complex because it needs new concepts and techniques, and new systems and management structures to create the right environment in which it can flourish.

The challenge is further complicated because DFM operates at different levels. At component level, decisions have to be made concerning ease, economy and quality of manufacture relating to choice of process. At assembly level, decisions need to be made on method of assembly technology and its configuration. And at product level, it is necessary to identify new product concepts that are compatible with existing or emerging manufacturing technologies.

Supporting and feeding into all of these levels are the principles, techniques and strategies for promoting DFM. An important message is that, for DFM to succeed, a company must apply the various components of DFM as a complete package, just as other modern approaches to manufacturing (for example, total quality management) have succeeded in transforming manufacturing activity. The Avery case study (Part 1) provided evidence to this argument. DFM cannot be practised effectively through a retrofitting of corrective action. The lessons and failures of value engineering should support this message. Effort has to be brought forward and put into the specification phase, where the concepts of the design are developed.

A basic problem with new approaches to manufacturing, like DFM, is that they can lack substance and detail, and accordingly ways to implement the necessary changes are difficult to identify and implement. In such situations, it can be argued, talking about DFM is much easier than practising it. In this sense, a textbook is limited in what it can achieve.

An aim of the book has been to show the breadth and depth of thought, the techniques and activity that support the whole subject. It is apparent from the range of material in the book and the issues raised therein that DFM affects many different people, ranging from component designers to product strategists. However, the bottom line is that designers at all levels must be more knowledgeable about modern manufacturing methods and technologies, and their implications for product design. But knowledge about manufacturability of particular components or product designs is not enough; it has to be translated into skills, so DFM can be practised.

DFM skills can be acquired in several ways. They can be obtained (perhaps artificially) through formal education coupled with an exposure to a real manufacturing situation with real manufacturing considerations. Alternatively, skills can be acquired more directly but explicitly through guidelines, checklists, the do's and don'ts of design for manufacture or assembly. A modern approach is to embed such manufacturing knowledge in computer-based methodology and present the information through the concept of a CAD/CAM system.

Early in the book, principles of DFM were discussed. These principles, as outlined in Part 2, take many forms, ranging from the multidisciplinary product development team through to checklists or general hints such as 'try to reduce the number of parts in order to eliminate assembly operations'. Such principles, it should be said, arise largely from experience based on lessons from applications where DFM has proved successful. The multi-disciplinary team approach, for example, has generally proved an effective way of bringing together product design and manufacturing specialists, and is practised by many companies. Such team work, to be successful, has to be driven from the top.

A valuable lesson or principle to be learned can be drawn from value engineering, a forerunner to DFM. There was nothing wrong with value engineering in itself; indeed, it encompasses many of the principles identified as being important to DFM. The problem lay in how value engineering was applied which, similarly to the concept of quality control as it used to be practised, meant that it was positioned outside the design to manufacture cycle and was not given the prominence it deserved. Also, because value engineering was not given sufficient status by being brought into mainstream activity, it focused on minimizing component costs rather than challenging the whole design concept, where more significant savings can be made.

In other situations, applying general principles to particular instances is not always appropriate. Design activity in its broadest sense is often a matter of compromise, in which design objectives have to be balanced against each other. In each instance, a set of objectives needs to be judged individually in relation to considerations of manufacture. This was clearly demonstrated by the General Motors transmission case in which a one-piece casting was replaced with three parts, going against the rule of reducing parts count. It did, however, on balance, benefit in other ways – from improved manufacturing processes.

Developments in manufacturing processes can of course lead to savings in production costs and reductions in production time. 'One-hit' machining set-ups and advances in casting technology clearly demonstrate such benefits, provided that products are appropriately designed at the outset to match the new manufacturing techniques. In a similar vein, it is important that manufacturing costs are also made available during the development of design concepts, and included in the specification alongside functional requirements. With accurate costings, it will be possible to balance a cost programme against product performance.

Rational methods, which encourage a systematic approach to DFM, were covered in Parts 3 and 4. Evaluation methods are aimed at providing, in a more formal manner, information on manufacturability to designers so they can choose between alternative configurations of design. Manufacturability information resulting from evaluation methods can be presented as checklists to guide a designer into making correct choices, given the costs of particular assembly processes. It should be remembered that, ideally, such information should be used creatively by the designer to improve the design.

Another approach to systematizing the DFM activity takes evaluation one step further and positions it within the wider framework of a CAD/CAM system. This has the effect of pushing DFM considerations firmly up towards the conceptual design of products, and hence DFM becomes an indispensable part of the total design process. This is more in line with concurrent engineering concepts currently being developed and applied in some industries.

More formal structures are employed in computer-aided DFM systems for encoding knowledge on manufacturability. Technologies to support this approach range from database technology, extensions of conventional geometrical models used in CAD systems, and those based on knowledge engineering and expert systems.

The dynamics of product and process design were discussed explicitly in Part 5. This is the 'top–down' strategic view of design and manufacturing operations, and lies in the business domain. It stresses the need to rethink traditional practices, and compares 'conventional' and 'new' paradigms. The important point is made that process and product management impact on each other in ways that are crucial to a company's strategy.

The role of top management is also crucial, as without its involvement, there is a strong danger that the integrated and simultaneous approach will disintegrate, with members aligning themselves with their original functional groups. The principles of DFM are simple enough in theory, but their practical implementation is much more difficult, requiring strong management participation. However, companies who have driven programmes from the 'top' are finding the rewards to be worthwhile in programmes which require fewer design changes, and have a bottom line of lower cost and higher quality.

Index